Research Methods for Pharmaceutical Practice and Policy

Introduction to the Pharmacy Business Administration Series

Books in the Pharmacy Business Administration Series have been prepared for use in university level graduate and professional level courses, as well as for continuing education and self-study uses. The series includes books covering the major subject areas taught in Social and Administrative Pharmacy, Pharmacy Administration, and Pharmacy MBA programs.

World-class authors with well-regarded expertise in the various respective areas have been selected and the book outlines as well as the books themselves have been reviewed by a number of other experts in the field. The result of this effort is a new integrated and coordinated series of books that is up to date in methodology, research findings, terminology, and contemporary trends and practices.

This is one book in that series of about 12 subjects in total. It is intended that each of the books will be revised at least every 5 years. Although the books were intended for the North American market, they are just as relevant in other areas.

Titles in the series currently include:
Health Economics
Health Policy and Ethics
Principles of Good Clinical Practice
Pharmaceutical Marketing: A Practical Guide
Financial Analysis in Pharmacy Practice

The series editor-in-chief is Professor Albert Wertheimer, PhD, MBA, of Temple University School of Pharmacy, Philadelphia

Suggestions and comments from readers are most welcome and should be sent to Commissioning Editor, Pharmaceutical Press, 1 Lambeth High Street, London SE1 7JN, UK.

Research Methods for Pharmaceutical Practice and Policy

Edited by
Rajender R Aparasu MPharm, PhD

Professor and Division Head of Pharmacy Administration and Public Health,
College of Pharmacy, University of Houston, Houston, Texas, USA

Pharmaceutical Press

Published by the Pharmaceutical Press

66-68 East Smithfield, London E1W 1AW, UK

First published 2011

Typeset by Thomson Digital, Noida, India
Printed in Great Britain by TJ International, Padstow, Cornwall

ISBN 978 0 85369 880 7

Dedication

To my parents, Surender and Anusuya Aparasu, and to my wife, Anu Aparasu

Contents

Preface

In recent years, pharmaceutical practice and policy research has gained prominence due to the increasing visibility of pharmacists, pharmaceuticals, pharmacist services, and pharmacy systems in the healthcare system. Providers, patients, payers, and policy makers are concerned about cost, access, and quality of pharmaceutical care. There is increasing pressure on pharmacists and pharmacy systems to improve efficiency, effectiveness, and safety of pharmaceuticals and pharmacist services. This requires evidence-based practices and policies rooted in sound scientific principles to achieve the desired healthcare goals.

Scientific innovations have played a prominent role in the growth of the pharmaceutical market. Pharmaceutical firms have traditionally funded basic and applied research on pharmaceuticals. Federal agencies such as the National Institutes of Health (NIH) and the Agency for Healthcare Research and Quality (AHRQ) have strongly supported safety and effectiveness studies for pharmaceutical products. These funding mechanisms have led to extensive growth in the pharmaceutical product knowledge base in terms of safety and effectiveness. It is not only vital to have a scientific knowledge base but also important to incorporate the existing knowledge into current practices and policies. This is often referred to as evidence-based practice and policy. Increasingly evidence-based medicine has become synonymous with quality of health. Several nonfederal agencies, such as the Joint Commission and National Committee for Quality Assurance, have developed process measures of quality from these proven practices. The AHRQ (2008) has funded several centers across the nation to generate and promote evidence-based practices and the emphasis on their implementation continues to increase.

Recently, researchers and practitioners are focusing on comparative effectiveness to improve efficient use of pharmaceutical products. The goal of the comparative research is to provide information to decision makers at both individual and population levels (Institute of Medicine 2009). The comparative effectiveness research has come to the forefront due to limited comparative data on safety and effectiveness. The efficacy data derived from placebo-controlled clinical trials are designed for the drug approval process.

Comparative effectiveness research is based on the concepts of evaluation of alternatives to select an appropriate agent to optimize patient outcomes. In fact, the AHRQ (2009) has funded the Centers for Education and Research on Therapeutics (CERTs) to conduct research to optimize use of medications.

The growth of the knowledge base and funding support from federal and nonfederal agencies clearly demonstrates the need and importance of pharmaceuticals in healthcare. However, the knowledge base and support for pharmacist services or pharmacy systems research has been limited when compared with other professional services and pharmaceutical product-based research. Concepts such as pharmaceutical care and medication therapy management need a strong knowledge base and a wider acceptance to implement evidence-based patient care practices. This requires concerted effort to move from abstraction to operationalization and evaluation. There is a need to develop a knowledge base that has internal validity (efficacy) and external validity (effectiveness). The effectiveness research has the additional benefit of providing generalizable information for policy making (Wells and Strum, 1996). With increasing emphasis on efficiency, effectiveness, and safety, the scientific knowledge base will be a critical component of pharmaceutical practice and policy.

Pharmaceutical practice and policy research has evolved over the years consistent with the changes in the profession, delivery of pharmaceutical care, and applied research methodologies. These applied research methodologies are derived from various disciplines. Consequently, faculty and students involved in pharmaceutical practice and policy research have traditionally used research methods textbooks from education, psychology, behavior sciences, and other applied research areas. Although there are several textbooks relevant to health services research, very few are available in pharmaceutical practice and policy research. This book is an earnest attempt to bring relevant content and expertise to conducting pharmaceutical practice and policy research. Its structure and content is based on years of experience teaching and applying the principles of pharmaceutical practice and policy research.

This textbook covers a wide range of topics for conducting scientific research starting with conceptualizing research and ending with statistical analysis. Traditionally graduate courses emphasize the theoretical aspect of applied research to provide a strong research foundation. However, this can lead to a strong understanding of concepts without appreciation for application. Pharmaceutical practice and policy research is applied research to address issues related to pharmacists, pharmaceuticals, pharmacist services, and pharmacy systems. The goal of the book is to provide both a theoretical and a practical framework for conducting pharmaceutical practice and policy research. Special emphasis is placed on examples and practical procedures that are important in implementing research.

There are 16 chapters in this textbook. Each chapter is designed to provide an understanding and application of the critical elements for conducting pharmaceutical practice and policy research. Chapter 1 defines pharmaceutical practice and policy research in the context of health services research, and explains the process of scientific inquiry in general and the overall process of pharmaceutical practice and policy research in particular. Chapter 2 explains concepts and constructs, and discusses the testing and application of common theories relevant to pharmaceutical practice and policy research. Chapter 3 focuses on the operationalization process, with emphasis on developing research problems, research questions, hypotheses, and operational definitions. Chapter 4 describes the concept and levels of measurement, discusses reliability and validity, and methods commonly used to evaluate reliability and validity. Chapters 5 and 6 provide experimental and nonexperimental approaches for conducting pharmaceutical practice and policy research. Chapter 5 discusses various types of experimental research designs to address cause and effect relationships and issues related to validity threats and Chapter 6 provides an overview of observational designs and techniques relevant for conducting nonexperimental research. Chapter 7 presents the key steps in designing a sampling plan including target population definition, sampling approaches, and determinants of sample size.

Chapter 8 provides a practical guide to understanding and implementing systematic reviews to uncover the existing knowledge base by ordering and evaluating the available literature. Chapters 9 and 10 discuss issues related to research methodologies in pharmaceutical practice and policy research. Chapter 9 provides an overview of qualitative, quantitative, and mixed data collection methods and Chapter 10 examines principles of survey design to provide a practical framework for designing and conducting effective surveys. Chapter 11 provides the statistical framework to analyze data for various research designs and questions. Chapters 12–14 describe research approaches based on secondary data: Chapter 12 provides the analytical framework including analytical files, data structure, and medical coding conventions to conduct research based on Medicare and Medicaid data; Chapter 13 provides an overview of the major types of commercial data and discusses issues to be considered when choosing commercial data sources; and Chapter 14 provides descriptions of common national survey data sources and research and analytical considerations in using national survey data for research. Chapter 15 describes approaches and reviews standards for conducting effective program evaluations. Finally, Chapter 16 examines major issues and trends influencing the future of pharmaceutical practice and policy.

This textbook is designed for entry-level graduate students in pharmacy administration, pharmaceutical evaluation or policy, or health services research focusing on pharmaceutical policy and practice issues. The contents of the book can be delivered in one semester. Each chapter builds on the previous chapters to logically progress from conceptualization to implementation of research studies. Additional reading and a project/proposal can improve the critical thinking skills of the students and also provide a mechanism to implement the research techniques. The online resources and chapter references can be used to supplement the content. In addition, recent journal articles relevant to the chapter can be useful in strengthening the conceptual and practical understanding of principles and techniques of conducting pharmaceutical practice and policy research. This book can also be an excellent resource for pharmacy students in professional programs, undergraduate students in pharmaceutical sciences, and pharmacists in residencies and fellowships, while also providing a useful tool to pharmacy practitioners and researchers.

I would really appreciate feedback from students and faculty for future editions. All knowledge is considered as work in progress including the contents of this book.

Rajender R Aparasu, MPharm, PhD
Houston, Texas
August 2010

References

Agency for Healthcare Research and Quality (2008). *Evidence-based Practice Centers: Synthesizing scientific evidence to improve quality and effectiveness in health care.* Overview. Rockville, MD: Agency for Healthcare Research and Quality. Online (updated November 10, 2008). Available at www.ahrq.gov/clinic/epc (accessed October 14, 2009).

Agency for Healthcare Research and Quality (2009). *Centers for Education & Research on Therapeutics (CERTs).* Rockville, MD: Agency for Healthcare Research and Quality. Online (updated March 6, 2009). Available at www.certs.hhs.gov (accessed October 14, 2009).

Institute of Medicine (2009). *Initial national priorities for comparative effectiveness research.* Washington, DC: National Academies Pr.

Wells KB, Strum R (1996). Informing the policy process: from efficacy to effectiveness data on pharmacotherapy. *J Consulting Clin Psychol* 64: 638–45.

Acknowledgments

I take this opportunity to thank my teachers, colleagues, and students who have influenced and challenged me throughout my career. In particular, I would like to extend my sincere thanks to two people who were the initial driving force for undertaking this endeavor, Drs Robert Mikeal and Albert Wertheimer. I sincerely appreciate their encouragement and support in developing the master plan for the book. I want to thank all the chapter contributors who worked diligently with me in developing the content for the book. I also thank the reviewers for their input to improve the content and flow of each of the chapters.

Appreciation is also in place for my graduate students for offering input from a student's perspective on several of the chapters. I would like specially to thank Mrs Golda Hallett for her help in all stages of the book's preparation. I am sure I would not have completed this book on time without her valuable assistance. Finally, I am grateful to Pharmaceutical Press and their staff, Ms Christina De Bono, Ms Lindsey Fountain, and Ms Rebecca Perry, for their help and support. The editorial assistance of Jane Sugarman and Marion Edsall is also very much appreciated.

About the editor

Rajender R Aparasu, MPharm, PhD, is a Professor and Division Head of Pharmacy Administration and Public Health at the University of Houston College of Pharmacy. Dr. Aparasu has over 15 years of experience in teaching and researching in the areas of pharmaceutical practice and policy. He has received several federal and nonfederal grants to address a wide variety of quality of pharmaceutical care issues. Dr. Aparasu has served as an expert consultant on Federal Patient Safety and APhA Medication Therapy Management Taskforces. He serves on the editorial boards of three peer-reviewed journals, *Research in Social and Administrative Pharmacy*, *The Journal of Drug, Healthcare and Patient Safety*, and *The Open Health Services and Policy Journal*. He is a peer-reviewer for several pharmacy and medical journals and has been recognized as an exceptional peer reviewer by *Medical Care* and *Pharmacoepidemiology and Drug Safety*. He is also an associate editor of *BMC Geriatrics*. Dr. Aparasu has made over 90 presentations in national and international meetings and has over 45 publications.

Contributors

Rajender R Aparasu, MPharm, PhD
Professor and Division Head, Pharmacy Administration and Public Health, Department of Clinical Sciences and Administration, College of Pharmacy, University of Houston, Houston, TX, USA

Darren Ashcroft, BPharm, MSc, PhD, MRPharmS
Reader in Medicines Usage and Safety and Director, Centre for Innovation in Practice, School of Pharmacy and Pharmaceutical Sciences, Manchester, UK

John P Bentley, PhD
Associate Professor of Pharmacy Administration and Research Associate Professor, Research Institute of Pharmaceutical Sciences, School of Pharmacy, University of Mississippi, MS, USA

Carolyn M Brown, PhD
Professor, Division of Pharmacy Administration, College of Pharmacy, The University of Texas at Austin, Austin, TX, USA

Richard R Cline, PhD
Associate Professor, Department of Pharmaceutical Care and Health Services, College of Pharmacy, University of Minnesota, Minneapolis, MN, USA

Betsey Jackson, MS
President, Health Data Services Corporation, Carlisle, MA, USA

Michael L Johnson, PhD
Associate Professor, Department of Clinical Sciences and Administration, College of Pharmacy, University of Houston, Houston, TX, USA

Khalid M Kamal, PhD
Assistant Professor, Division of Clinical, Social and Administrative Sciences, Mylan School of Pharmacy, Duquesne University, Pittsburgh, PA, USA

Kenneth A Lawson, PhD
Associate Professor and Mannino Fellow, Pharmacy Administration Division, College of Pharmacy, The University of Texas, Austin, TX, USA

Bradley C Martin, PharmD, PhD
Professor and Director of Pharmaceutical Evaluation and Policy, Division of Pharmaceutical Evaluation and Policy Faculty, University of Arkansas for Medical Sciences, Little Rock, AR, USA

Jon C Schommer, PhD
Professor and Associate Department Head, Department of Pharmaceutical Care and Health Systems and Director of Graduate Studies in Social and Administrative Pharmacy Track, College of Pharmacy, University of Minnesota, Minneapolis, MN, USA

Robert J Valuck, PhD
Associate Professor, Department of Clinical Pharmacy, School of Pharmacy, University of Colorado Denver, Aurora, CO, USA

Albert I Wertheimer, PhD
Professor of Pharmacy, School of Pharmacy, Temple University, Philadelphia, PA, USA

Marcia M Worley, PhD, RPh
Associate Professor, Department of Pharmacy Practice and Pharmaceutical Sciences, College of Pharmacy, University of Minnesota at Duluth, Duluth, MN, USA

1

Scientific approach to pharmaceutical practice and policy research

Rajender R Aparasu

Chapter objectives

- To define pharmaceutical practice and policy research
- To introduce the knowledge base and stakeholders in pharmaceutical policy research
- To describe the guiding principles for scientific inquiries
- To discuss the stages of pharmaceutical practice and policy research

Introduction

Pharmaceuticals and pharmacist services are integral parts of healthcare delivery, and pharmacists and pharmacy systems play a critical role in current healthcare. Pharmaceutical practice and policy research is an applied field of research that deals with issues impacting pharmaceuticals, pharmacists, and pharmacy systems. It has evolved over the decades consistent with changes in healthcare delivery in general and the evolution of pharmaceutical care in specific. Applied research methodologies in health services research have also been instrumental in the growth of pharmaceutical practice and policy research. Pharmaceutical practice and policy research helps in the growth of the applied scientific knowledge base to aid decisions relevant to patients, providers, payers, and policy makers. It also helps to improve the practice of pharmacy and promote the profession. With increasing complexities of delivering pharmaceutical care, research has become the cornerstone in safe and effective delivery of pharmaceutical care.

The goals of health and pharmaceutical care are centered on improving populations' health and quality of care (Hepler and Strand 1990; Shi and Singh 2004; Institute of Medicine 2008). This requires that care of patients be grounded in scientific research (Institute of Medicine 2008). Pharmaceutical practice and policy research will not only help to improve the quality of care but also enhance the role of pharmacists and pharmacy systems in the delivery of healthcare. Health policy, including policies directed toward pharmaceutical and pharmacy systems, involves actions by governmental agencies and other players designed to influence the health of the population (Miller 1987; Ruwaard et al. 1994). These administrative actions influence the distribution and utilization of healthcare resources and delivery. To this end, research can be instrumental in the development and implementation of pharmaceutical policies (Willison and MacLeod 1999). This chapter defines pharmaceutical practice and policy research in the context of health services research, and presents the knowledge base and stakeholders in pharmaceutical practice and policy research. Finally this chapter explains the process of scientific inquiry in general and the process of pharmaceutical practice and policy research in particular.

Pharmaceutical practice and policy research

Health services research

Health services research is the broadest applied field of research dealing with all aspects of the healthcare system. Pharmaceutical practice and policy research is a component of health services research dealing with the delivery of pharmaceuticals and pharmacist services. An understanding of health services research can help to appreciate the scope and the goals of pharmaceutical practice and policy research. The Academy for Health Services Research and Health Policy, also referred to as AcademyHealth, has defined health services research as a "multidisciplinary field of scientific investigation that studies how social factors, financing systems, organizational structures and processes, health technologies, and personal behaviors affect access to health care, the quality and cost of health care, and ultimately our health and well-being" (Lohr and Steinwachs 2002). According to the Agency for Healthcare Research and Quality (AHRQ 2002), health services research examines how people access healthcare, the cost of care, and the result of this care on patients. The Institute of Medicine (IOM 1994) has defined health services research as "a multi-disciplinary field of inquiry, both basic and applied, that examines access to, and the use, costs, quality, delivery, organization, financing, and outcomes of health care services to produce new knowledge about the structure, processes, and effects of health services for individuals and populations." These definitions emphasize the scope and breadth of health services research in addressing cost, access, and quality of healthcare.

According to the IOM (1991), the goal of health services research is "to provide information that will eventually lead to improvements in the health of the citizenry." The research domains of health services research are individuals, families, organizations, institutions, communities, and populations (Lohr and Steinwachs 2002). The stakeholders who have an interest in health services research include patients, providers, payers, healthcare administrators, and policy makers. The goals of health services research are more specific, according to the AHRQ (2002). The main goals are "to identify the most effective ways to organize, manage, finance, and deliver high quality care; reduce medical errors; and improve patient safety." These goals focus on generating knowledge to improve efficiency, effectiveness, and safety of healthcare services.

The IOM (1991) provided broad direction in the areas of health services research to address the future healthcare needs, including the following:

- *Organization and financing of health services*: healthcare delivery and financing are changing due to increasing healthcare expenditures. Research can evaluate the impact of health insurance and cost containment measures on access, cost, and quality of healthcare.
- *Clinical evaluation and outcomes research*: assessment and improvement of quality of care are paramount in delivery of healthcare. Research can provide evidence-based data on prevention, diagnoses, and treatment to facilitate rational and effective decisions.
- *Monitoring and accountability*: clinical and financial responsibilities have increased for providers, payers, and policy makers. Research can develop and evaluate well-accepted clinical and financial indicators based on the methodologies derived from operational research and industrial management.
- *Informatics and clinical decision-making*: medical information systems will play a key role in the delivery and evaluation of healthcare services. Research can document and disseminate the role of informatics in clinical decision-making.
- *Populations and communities*: healthcare delivery systems serve the healthcare needs of populations and communities that are ethnically and culturally diverse. Research can help to understand disparities in race, and patient behaviors that impact the health of individuals and populations.
- *Provider and consumer behavior*: consumers, policy makers, managers, and providers are increasingly making decisions within internal and external constraints. Research can help to examine clinical, ethical, and financial considerations in prevention, diagnoses, and treatment.

- *Healthcare personnel*: diverse professionals with varying training and background are involved in the delivery of complex medical care. Research can address the training and work force needs of physicians and allied healthcare personnel to improve access, cost, and quality of healthcare.

Pharmaceutical practice and policy research

Historically, the term "pharmacy administration" has been used to define pharmaceutical practice and policy area because it mostly dealt with administration issues, such as acquisition, management, and operations in pharmacy (Blaugh and Webster 1952). The areas covered included accounting, economics, marketing, management, and law. Manasse and Rucker (1984) defined pharmacy administration as a "sub-discipline of the pharmaceutical sciences which centers on the study of and education in the applied social, behavioral, administrative and legal sciences which bears upon the nature and impact of pharmacy practice regardless of environment." Hepler (1987) proposed the following research areas for support of pharmacy practice: relationship and organization of the pharmacy profession in society, professionalization including training and evaluation, drug use control involving formative and summative research, and organization decision-making.

Einarson (1988) defined pharmacy administration as "an applied discipline involving the study of phenomena and relationships associated with drugs." It encompasses research, evaluation, management of pharmaceuticals, patients, and pharmacy systems. According to Einarson, pharmacy administration issues are examined at "micro, macro, and global levels from financial, economic, managerial, legal, ethical, and social, behavioral, educational, and historical perspectives." Einarson's definition is consistent with the definition of health services research but is specific to pharmacy issues. In the last few decades, practice- and policy-based research has expanded from community pharmacy issues to system level issues covering broad areas related to delivery of pharmaceuticals, pharmacist services, and pharmacy systems.

The proposed definition extends Einarson's (1988) definition within the context of health services research. For the purposes of this book, pharmaceutical practice and policy research is defined as a multidisciplinary field of scientific investigation that examines cost, access, and quality of pharmaceutical care from clinical, sociobehavioral, economic, organizational, and technological perspectives. It increases knowledge and understanding of the structure, processes, and outcomes of pharmaceuticals, pharmacist services, and pharmacy systems for individuals and populations. Understanding pharmaceutical practice and policy research in the context of health services research provides the subject breadth and methodological depth to conduct research to address practice- and policy-related issues across the healthcare spectrum. Pharmaceutical practice and policy research terms and definitions reflect the changes in the practice of pharmaceutical care and advances made in areas such as pharmacoepidemiology and pharmacoeconomics. The goal of pharmaceutical practice and policy research is to provide information on pharmaceuticals and pharmacist services to improve patient outcomes. Pharmacists and pharmacy systems are the means by which the practices and policies based on scientific research can be implemented.

Pharmaceutical care practices have evolved due to complexities in delivering pharmaceuticals and pharmacist services in diverse healthcare settings. The current stakeholders such as investors, healthcare institutions, prescribers, payers, patients, regulators, and policy makers reflect these complexities. The prominence of chain pharmacies in the market place has altered the dynamics of prescription delivery and pharmacist services in community settings. Health-system pharmacies have developed diverse medication delivery systems and medication therapy management services specific to institutional settings. The role of allied health professionals has expanded in the area of prescription medications; no longer is the physician the sole prescriber of medication. Insurance companies and pharmacy benefit managers, involved in reimbursing and managing pharmacy benefits, have created diverse prescription plans to deal with rising prescription costs. Medicaid and Medicare are increasingly playing a role in pharmacy reimbursement systems for medications and medication therapy management programs. The role of policy makers and regulators gained prominence due to increasing visibility of pharmaceuticals, pharmacists, and pharmacy systems in healthcare. The patient's role in the medication use process is still prominent due to cost shifting, increasing access to health information, and growth of prescription alternatives.

The same knowledge base is necessary to conduct both pharmaceutical practice and policy research and health services research. Awareness of research concerns in each of these areas is vital in conducting pharmaceutical practice and policy research. Consequently, the methodologies used are as diverse as the issues. The research methodologies are often derived from epidemiology, economics, sociology, psychology, statistics, management, and marketing. The research techniques vary based on the expertise, availability of tools, and objectives of the study; the key elements of scientific inquiry have, however, remained the same over the years. Understanding the key elements of scientific inquiry is critical in conducting pharmaceutical practice and policy research.

Scientific inquiry

The purpose of scientific inquiry is to create knowledge or a body of information that can explain observed phenomena. The information generated from scientific inquiry is used to develop theory or laws. Although some may argue that theory and laws have different connotations based on the acceptance of knowledge, both emanate from the findings of scientific inquiries. Frequently, theories and laws are tested scientifically to confirm their applicability to observed phenomena across research areas, and may then be revised to reflect a new body of knowledge. Theories or models are often used in health services research. Theories commonly discussed in pharmaceutical practice and policy research include the Andersen Model of Health Care Utilization Behavior, Health Belief Model, and Theory of Planned Behavior (Ajzen 1991; Janz et al. 2002; Andersen 2008).

Scientific research is defined as a "systematic, controlled, empirical, and critical investigation guided by theory and hypothesis about presumed relationships among such phenomena" (Kerlinger 1986). Research is an ongoing process to refine the existing knowledge of relationships. Knowledge generated through research is used by practitioners and policy makers to make decisions to improve the health of the population. The term "evidence-based practice" is currently being used to strengthen healthcare practices based on research (IOM 2008). As research is an ongoing process it is critical to make healthcare decisions based on the current state of knowledge. Often, lack of appreciation and application of the key elements of scientific inquiry can endanger practices and policies (Majumdar and Soumerai 2009). Sometimes adoption of knowledge generated by scientific research for practice as well as policy lags behind for various reasons. In recent years concerted effort is being made to implement evidence-based practice and policy at various levels of healthcare delivery.

The National Research Council (NRC 2002) developed guiding principles for all scientific inquiries. These principles are applicable to any scientific research, including pharmaceutical practice and policy research. The guiding principles purport that the research should: (1) pose a significant scientific question that needs empirical investigation; (2) link research inquiry to relevant theory; (3) utilize relevant methodologies that involve direct investigation; (4) provide coherent and explicit reasoning; (5) replicate and generalize across studies; and (6) provide a mechanism for scientific scrutiny. These principles strengthen the process of scientific inquiry and help to generate knowledge that benefits society.

Pose a significant scientific question that needs empirical investigation

This principle requires two characteristics, namely significance and empiricism. The significance of a question is judged on the quality of the question in relation to the existing knowledge base and information generated by research to advance the existing knowledge. The significance of a question can be evaluated based on the literature review on the topic and its relevance to practice and policy. Pharmaceutical practice and policy research is considered to be applied research; consequently, the research question should have clear implications for practice and policy. Empiricism refers to information obtained by experience or explicit observation. It is based on the philosophy of positivism (Green and Browne 2005). Positivism is knowledge of natural phenomena derived from empirical evidence. Empirical data necessitate conceptualization and measurement; therefore, the scientific research questions must possess these characteristics. The objectivity and verifiability of empirical data are critical to research. Objectivity ensures that there is no bias in the data used for the research.

Link research inquiry to relevant theory

Scientific theories are valuable in understanding the research question under an existing or newly developed relationship framework. They are conceptual models used to observe and explain a phenomenon. Theories help to make the phenomenon of interest understandable in a broad conceptual, as well as a practical, framework (Brazil et al. 2005). The conceptual framework, based on a single theory or multiple theories, elucidates the relationships of interest and guides the hypothesis development in the research process. It also provides a framework for other aspects of the research process, including data collection, analysis, and interpretation. Often, pharmaceutical practice and policy research is based on behavioral, economic, social, psychological, and medical theories. Although theory-driven research is important in the research process, it may not be possible to link all research questions to a theory in pharmaceutical practice and policy research. Sometimes, the knowledge base in a given area of interest may not permit utilization of theory as theories are not available or not well developed in that specific area. Such studies are referred to as explorative studies. These studies are driven by a practically relevant question.

Utilize relevant methodologies that involve direct investigation

Scientific methods include research designs, data collection procedures, and analysis of data. Research designs refer to the structural framework of investigation (Mikeal 1980). They include specification of variables of interest, time order sequencing of variables, and sampling and allocation procedures. The primary focus of research is on stimulus and response variables. The stimulus variable is also referred to as the intervention, exposure, or independent variable. The response variable is also known as the outcome, disease, or dependent variable. These variables have to be clearly specified and operationalized. The time order sequencing refers to the timing of the measurement of variables of interest. The classification of study designs such as a cross-sectional or longitudinal design is based on the time order sequencing. Data collection procedures operationalize the research question and variables of interest. Empirical methodologies are based on acceptable definition and data collection methods. The data collection process can be self-reports, physiological measurement, or behavior observation. This can involve collection of data by the researcher, called primary data, or data collected by others for reuse, called secondary data. Data collection procedures are critical in the research process and they have to be planned and implemented properly to achieve the research objective. Data analysis involves processing of data based on statistical principles. The choice of statistical analysis is based on the research question, type of variables, research design, and other factors.

Provide coherent and explicit reasoning

Inferential reasoning involves explanations, conclusions, and predictions based on the research findings. It is also needed to explain the information obtained from data analysis. This requires an understanding of underlying theory, formulated hypothesis, research design and methodology, and research findings. The reasoning and explanations for linking the findings to the theory or previous research should be logical. Often, research guided by theory and hypothesis is strong due to their ability to explain the research findings (Brazil et al. 2005). The strength of the evidence is interpreted based on these explanations. The discussion of the findings also requires understanding of limitations. The limitations of research are driven by research design, data collection procedures, and analysis.

Replicate and generalize across studies

This principle addresses two interrelated concepts in sociobehavioral research, namely reproducibility and validity. Reproducibility, also known as reliability, refers to consistency of evidence across studies. Data collection and the measurement process in individual studies focus on reliability and validity to strengthen the study findings. The reproducibility principle emphasizes the reproducibility of empirical evidence from one research study to another. Consistency of findings strengthens the evidence as well as the underlying theory or model in applied health and pharmaceutical research. Applied research involves implementation of research methodologies in realistic settings. The extent of control in applied research is not the same as in the natural sciences. In addition, the inherent variability in research methodologies, such as surveys and interviews, requires reproducibility to strengthen the research findings. Generalization, also scientifically known as external validity, refers to the extent to which the findings are applicable to other people, places, and time. Applied health and pharmaceutical research is constrained for various reasons, including samples, location, duration, and resources. It is important to generalize the findings across time, space, and populations. This may also require further research of diverse people, places, and times. The continued discussion of efficacy and effectiveness is a classic example of the generalizability issue (Godwin et al. 2003). Efficacy evidence is derived from the data based on ideal and controlled clinical settings. Effectiveness evidence is derived from studies based on actual practice settings. The ability to generalize findings from clinical trials (efficacy) to actual practice (effectiveness) is an indicator of the strength of the evidence and its applicability across time, space, and populations.

Provide a mechanism for scientific scrutiny

Scientific research requires scrutiny by peers; this is critical for scientific progress. All aspects of implementation of research, research findings, reasoning, and conclusions must be carefully reviewed. Scientific scrutiny entails discussion and debate by peer researchers and practitioners. The research report, also known as a manuscript, is often submitted to a peer-reviewed journal for scientific scrutiny. All aspects of the research process are made explicit in the introduction, methods, results, and discussion sections of the manuscript. Manuscripts are often reviewed by two or more reviewers consisting of researchers and practitioners. The peer-review process is designed to provide constructive criticism to improve the quality of the research. Acceptable manuscripts are published in journals and add to the knowledge base of the profession, guiding practice and policy. In addition to publication, other venues for scientific scrutiny include meetings, conferences, and public policy discussions. However, the process may or may not be as explicit as peer-reviewed journals. If the principle of scientific scrutiny is disregarded, such research does not add to the knowledge base of the profession even though it may be based on all other scientific principles. With increasing emphasis on evidence-based practice and policy, there is a strong need to make the research findings available to researchers, practitioners, and policy makers via the publication process.

Stages of pharmaceutical practice and policy research

Pharmaceutical practice and policy research addresses diverse issues from different perspectives related to pharmaceuticals, pharmacists, and pharmacy systems. It may require consideration of research issues within a clinical, sociobehavioral, economic, organizational, or technological context. Regardless of the perspective, it is important to know the implications of research on pharmaceutical practice and policy. The knowledge base to understand the issues and implement the research may or may not be the same. Sometimes it requires application of research methods uncommon to pharmaceutical practice and policy research when addressing a research question. The research plan designed to address the research questions has to be based on the NRC's (2002) six principles of scientific research previously discussed. The guiding principles provide the framework to conduct pharmaceutical practice and policy research.

The stages of pharmaceutical practice and policy research can be grouped into: (1) research problem, (2) research hypothesis, (3) research design, (4) research methodology, (5) statistical analysis, and (6) research report. These stages are based on the six guiding principles of scientific research. Each stage provides a critical component of the research framework and can be considered interdependent as each stage provides the foundation for the next stage. The research problem determines the type of research hypothesis. Similarly, the research design is selected based on the research hypothesis. Although the stages are presented sequentially, actual implementation may or may not be sequential for practical and scientific reasons. Sometimes practical considerations in research methodologies may necessitate a change in research question. Brief descriptions of each stage are provided below and discussed at length in each of the chapters of this book.

Research problem

The research problem describes the problem to be solved (Kerlinger 1986). According to the scientific principles (NRC 2002), a research question requires significance and empiricism. The significance of a question can be evaluated for its relevance to pharmaceutical practice and policy based on the existing knowledge base. The question has to be evaluated empirically to make a contribution to current and future pharmaceutical practice and policy.

Research hypothesis

The research hypothesis is a statement of the expected relationship (Kerlinger 1986). An empirical research question can be valuable in formulating a research hypothesis. According to the scientific principles (NRC 2002), research inquiry should be linked to theory. Research based on the theory can be instrumental in developing the hypothesis.

Research design

Research designs should specify the structural framework of investigation including variables of interest, time order sequencing of variables, and sampling and allocation procedures (Mikeal 1980). The research hypothesis provides the framework for research design. The variables of interest include stimulus and response variables. The timing of the measurement of variables is based on the hypothesis. The time order sequencing is reflected in various research designs available for pharmaceutical practice and policy research. According to the scientific principles, relevant methodologies involve selection of appropriate research design to investigate a research question.

Research methodology

Research methodologies refer to the data collection procedures to address a research question. Pharmaceutical practice and policy research entails diverse data collection procedures that have been developed based on the research question. Research data can be primary or secondary data, or a combination. The primary data collection can involve observation, self-reports, or clinical evaluations. Secondary data can include medical charts, insurance claims, or national surveys. According to the scientific principles, data collection procedures should be based on relevant methodologies that involve direct investigation.

Statistical analysis

Statistical analysis involves application of the principles of probability and statistical concepts to analyze the research data. The goal of statistical analysis is to answer the research question based on the data collected. The nature and extent of statistical analysis are based on the research question, research hypothesis, research design, and research methods. According to the scientific principles, relevant methodologies involve selection of appropriate statistical analysis to analyze research data.

Research report

The research report is the end product of a research endeavor and it captures the explicit research process, including research question, research hypothesis, research design, research methods, and statistical analysis. It should provide coherent and logical reasoning for the findings. The research report should address the three principles of scientific research, namely coherent and explicit reasoning, replication and generalization, and scientific scrutiny.

This book is not designed to provide details on scientific writing because several books address such scientific writing at length (see Bonk 1998; Albert 2000). A brief discussion on the three principles of scientific research is, however, provided below in the context of writing a peer-reviewed journal article. Although there are other avenues to publish a research report, journal articles are often used to report research findings.

Most research articles in a peer-reviewed journal follow a sequence of sections, namely introduction, methods, results, and discussion (Zellmer 1981; Pakes 2001). Other elements of the research report such as abstracts and references vary from journal to journal. The guidelines for manuscript format may also vary. The introduction is the first major section of the research report. It generally includes background information on the research topic and should briefly cover the existing knowledge base in the research area and provide a rationale for the current research. This section also includes the research problem and research hypothesis. It should provide coherent and explicit reasoning for conducting the research. The methods section should provide all details of research design, data collection methods, and statistical tests used to achieve the research objective. This is important to ensure the explicit nature of scientific research. The goal of this section is to provide all relevant details of the study implementation so that the findings can be replicated. This also ensures that the readers have a clear understanding of the context of the research process when evaluating research findings.

The results section provides all relevant research findings. The presentation of findings should be structured on the basis of the research objective. The study sample is often described using summary statistics such as proportions and means. Often detailed data are presented in the form of tables and figures. Generally information presented in tables and figures is not duplicated in the text, only key findings being presented there. It is important to indicate the variability of the measures, such as standard deviation and confidence intervals. Findings related to statistical tests should include probability or p values to explicitly state the sampling error. In general, significant and nonsignificant findings are presented and judgment on the findings is not included in the results section. The results section is considered the most explicit and objective section of the research report.

The discussion section of the report provides interpretation and scientific reasoning for the findings. Coherent and explicit reasoning is an important scientific principle to explain research findings. The findings and the conclusions should be consistent with the research question, research hypothesis, research design, research methods, and statistical analysis. The core of the discussion should focus on findings related to the hypothesis. The scientific reasoning often relates to the theory and previous research. Some findings are designed to replicate or validate previous studies. This strengthens the arguments and the underlying theory. The discussion section is also used to address limitations of the study findings. Each of the components of the journal article is critical for communicating research findings and vital for scientific discourse and scientific scrutiny.

Summary and conclusions

Pharmaceutical practice and policy research addresses a wide variety of research areas dealing with pharmaceuticals, pharmacists, and pharmacy systems. It is an applied research designed to create knowledge or a body of information to improve efficiency, effectiveness, and safety of pharmaceuticals and pharmacist services. The research is vital for practitioners, payers, policy makers, and researchers involved in pharmaceutical care. Pharmaceutical practice and policy research is based on the following guiding principles for scientific inquiries: (1) to pose a significant scientific question that needs empirical investigation, (2) to link research inquiry to relevant theory, (3) to utilize relevant methodologies that involve direct investigation, (4) to provide coherent and explicit reasoning, (5) to replicate and generalize across studies, and (6) to provide a mechanism for scientific scrutiny. Based on these principles, the stages of pharmaceutical practice and policy research can be divided into research problem, research hypothesis, research design, research methodology, statistical analysis, and research report. Each stage provides the basis for the next stage and the strength of scientific research is based on the incorporation of guiding principles into each of these stages. With increasing emphasis on evidence-based practice and policy, there is a strong need to strengthen and expand the knowledge base of pharmaceutical practice and policy.

Review topics

1 Define health services research and pharmaceutical practice and policy research.
2 Discuss the six guiding principles for scientific inquiries.
3 Discuss the stages of pharmaceutical practice and policy research.
4 Describe the stakeholders of pharmaceutical practice and policy research.
5 Discuss any two areas of knowledge base to conduct pharmaceutical practice and policy research using examples.

References

Andersen RM (2008). National health surveys and the behavioral model of health services use. *Med Care* 46: 647–53.

Agency for Healthcare Research and Quality (2002). *What is Health Services Research?* Rockville, MD: Agency for Healthcare Research and Quality. AHRQ Publication No. 02-0011. Available at: www.ahrq.gov/about/whatis.htm.(accessed on October 15, 2009).

Ajzen I (1991). The theory of planned behavior. *Organiz Behav Human Decision Processes* 50: 179–211.

Albert T (2000). *Winning the Publications Game*, 2nd edn. Abingdon: Radcliffe Medical Press.

Blaugh LE, Webster GL (1952). *The Pharmaceutical Curriculum*. Washington, DC: American Council on Education.

Bonk RJ (1998). *Medical Writing in Drug Development. A practical guide for pharmaceutical research*. New York: Haworth Press Inc.

Brazil K, Ozer E, Cloutier MM, Levine R, Stryer D (2005). From theory to practice: improving the impact of health services research. *BMC Health Serv Res* 5: 1.

Einarson TR (1988). Clinical pharmacy administration. *Drug Intell Clin Pharm* 22: 903–5.

Godwin M, Ruhland L, Casson I, *et al.* (2003). Pragmatic controlled clinical trials in primary care: the struggle between external and internal validity. *BMC Med Res Methodol* 3: 28.

Green J, Browne J, eds (2005). *Principles of Social Research*. New York: McGraw-Hill.

Hepler CD (1987). Pharmacy administration and clinical practice research agenda. *Am J Pharm Educ* 51: 419–22.

Hepler CD, Strand LM (1990). Opportunities and responsibilities in pharmaceutical care. *Am J Hosp Pharm* 47: 533–43.

Institute of Medicine (1991). *Improving Information Services for Health Services Researchers: A report to the National Library of Medicine*. Washington, DC: National Academy Press.

Institute of Medicine (1994). *Health Services Research: Opportunities for an expanding field of inquiry – an interim statement*. Washington, DC: National Academy Press.

Institute of Medicine (2008). *Knowing what Works in Health Care: A roadmap for the nation*. Washington, DC: The National Academies Press.

Janz NK, Champion VL, Strecher VJ (2002). The health belief model. In: Glanz K, Rimer BK, Lewis FM (eds), *Health Behavior and Health Education: Theory, research and practice*, 3rd edn. California: Jossey-Bass, 45–66.

Kerlinger FN (1986). *Foundations of Behavioral Research*, 3rd edn. New York: Holt, Rinehart & Winston.

Lohr KN, Steinwachs DM (2002). Health services research: an evolving definition of the field. *Health Serv Res* 37(1): 7–9.

Majumdar SR, Soumerai SB (2009). The unhealthy state of health policy research. *Health Affairs* 28: w900–8.

Manasse HE, Rucker TD (1984). Pharmacy administration and its relationship to education and practice. *J Soc Admin Pharm* 2: 127–35.

Mikeal RL (1980). Research design: general designs. *Am J Hosp Pharm* 37: 541–8.

Miller CA (1987). Child health. In: Levine S, Lillienfeld A (eds), *Epidemiology and Health Policy*. New York: Tavistock Publications, 15.

National Research Council (2002). Committee on Scientific Principles for Education Research. In: Shavelson RJ, Towne L (eds), *Scientific Research in Education*. Center for Education. Division of Behavioral and Social Sciences and Education. Washington, DC: National Academy Press.

Pakes GE (2001). Writing manuscripts describing clinical trials: a guide for pharmacotherapeutic researchers. *Ann Pharmacother* 35: 770–9.

Ruwaard D, Kramers PGN, Berg Jeths A van den, Achterberg PW, eds (1994). *Public Health Status and Forecasts: The health status of the Dutch population over the period 1950–2020*. The Hague: SDU Uitgeverij.

Shi L, Singh DA (2004). *Delivering Health Care in America: A systems approach*, 3rd edn. Sudbury, MA: Jones & Bartlett.

Willison DJ, MacLeod SM (1999). The role of research evidence in pharmaceutical policy making: evidence when necessary but not necessarily evidence. *J Eval Clin Pract* 5: 243–9.

Zellmer WA (1981). How to write a research report for publication. *Am J Hosp Pharm* 38: 545–50.

Online resources

US National Library of Medicine. *Introduction to Health Services Research*. A Self-Study Course. Online. Available at: www.nlm.nih.gov/nichsr/ihcm/index.html.

US National Library of Medicine. National Information Center on Health Services Research and Health Care Technology (NICHSR). Available at: www.nlm.nih.gov/hsrinfo.

2

Conceptualizing research

Carolyn M Brown

Chapter objectives

- To present concepts and constructs in healthcare
- To discuss common theoretical models of healthcare utilization and outcomes
- To explain various types of variables in pharmaceutical practice and policy research
- To understand the scientific relationships in pharmaceutical practice and policy research

Introduction

A scientific investigation in pharmaceutical practice and policy research begins with the conceptualization phase. Conceptualization refers to the process of developing and refining theoretical ideas. In quantitative research, it is basically a process of finding appropriate theoretical constructs for explaining or understanding practical phenomena. One primary step in the conceptualization process involves the conceptual or theoretical framework for the study. Whether the goal is to modify patient or provider behavior or to evaluate pharmacy services, the use of theory is critical to providing a coherent, systematic view of the phenomenon beyond the specifics of the study at hand. Thus, theory-driven research is more generalizable and its findings are more relevant to practice and policy developments. The purpose of this chapter is to demonstrate the importance of theory-guided research and how the use of theoretical frameworks is needed in and useful to pharmaceutical practice and policy.

In this chapter, the definitions of concepts and theory are presented, followed by a description of the testing and application of theory. Next, common theoretical models and their application in pharmaceutical practice and policy research are presented, followed by a discussion of types of variables and scientific relationships in pharmaceutical practice and policy research.

Definition of concepts and constructs

A concept is a word or term that stands for something – a feeling, a characteristic, a behavior, or an object. Concepts are abstractions of particular feelings, characteristics, or behaviors. For example, what is the meaning of the concept "healthcare quality?" It could be the absence of medication dispensing errors or a lower incidence of preventable drug-related problems. In either case, healthcare quality is abstracted from a particular behavior or characteristic. Other common concepts in healthcare include health outcomes, access to care, and health behaviors. A construct, on the other hand, is a concept with the added meaning of having been constructed for purposes of research (Kerlinger 1986). In practice, terms such as concept, construct, and variable are used interchangeably, although theoretically they have different meanings.

In order for concepts (constructs), which are non-observables, to be useful in research studies, they must be operationally defined so that they can be measured. For example, access to care has been defined in terms of physician visits, presence of health insurance, and delays in seeking care. In each case, access to care was translated into a variable that can be measured in an empirical investigation. Variables and definitions are used to create observable (measurable) forms of abstract concepts and constructs. Simon (1978) succinctly captured this idea when he noted that, "Operationalization of the theoretical concepts is the task of finding appropriate empirical proxies for the theoretical variables. Conceptualization is the complementary process of finding appropriate theoretical constructs for interesting empirical patterns that turn up."

Concepts are the building blocks of theory and express the abstract notions within a theory. When designing studies to investigate phenomena in pharmaceutical practice and policy, there are many conceptual issues to consider and the use of theory or theoretical frameworks is central to this conceptualization process.

Theory in healthcare research

Definition of theory

The primary goal of science is theory, namely, explaining natural phenomena (Kerlinger 1986). Kerlinger (1986) defines theory as "a set of interrelated constructs (concepts), definitions, and propositions that present a systematic view of phenomena by specifying relations among variables, with the purpose of explaining and predicting phenomena." These propositions are statements about the relationship between variables, i.e., how change in one variable is related to change in another variable. A theoretical model or framework is more like a "mini-theory" because it focuses on a few elements abstracted from all of reality and is not necessarily as well "worked out" as a theory (Simon 1978). However, the terms "theoretical model" and "theory" are commonly used interchangeably.

A theoretical model guides research, determining what will be measured and what statistical relationships will be proposed and tested. It serves as a foundation or roadmap for the research investigation by providing the key constructs hypothesized to influence the phenomenon of interest. It provides focus and boundaries for research studies and has been described as a double-edged sword (Crosby et al. 2002). On the one hand, the boundaries provide structure and focus to research investigations. On the other, the same boundaries can also serve as a limitation because potential other important variables, not included in the theory, are not examined. Although multiple theoretical frameworks are available, with many reflecting similar ideas, each framework offers unique constructs that are thought to be important. Thus, pharmaceutical practice and policy researchers should examine research problems from various perspectives and use multiple frameworks in order to uncover relationships that may otherwise go undiscovered.

In pharmaceutical practice and policy research, many theoretical models are employed. However, not all studies will employ an explicit theoretical model, but all studies have a theoretical basis. Often, simple or specific relationships (such as between race/ethnicity and medication adherence) are investigated, but those are not as generalizable or as well understood as relationships that are part of an integrated theoretical framework of individual decision-making (such as the Theory of Planned Behavior). Theoretical relationships are more general and widely applicable (Kerlinger 1986; Brazil et al. 2005).

Theories can be thought of as efficient ways to examine phenomena because they are general and can be applied to many specific cases. Theories intentionally transcend a particular time, place, group of people, or situation (Polit and Beck 2008). Instead of developing separate theories to explain each of the various types of individual behaviors (e.g., medication adherence, glucose monitoring), a general theory of individual health behavior is developed to explain the different types of individual behavior, including patient health behaviors. For example, the Theory of Planned Behavior (discussed later in the chapter) can be used to study disparate behaviors such as medication adherence, exercise behaviors, and condom use. Theories are built inductively from research findings, logical thought, and practical experiences, and the hypotheses that are deduced from theories, in turn, must undergo empirical testing. Theory building and theory testing are a dynamic processes and form the basis for scientific inquiry.

Testing of a theory

Theories are not directly tested, but the hypotheses that are formulated deductively from the theory are empirically tested. A hypothesis is a statement about the expected relationships between two or more variables. Hypotheses serve as bridges between theory and the empirical study (Kerlinger 1986) and they should flow directly from the theoretical model or framework. The concepts or variables included in the hypothesis must be relevant to the research question and the hypothesis must be testable. Hypotheses are never really proved or disproved. Rather, research findings may support or not support the hypothesis (Simon 1978). Support for theoretically derived hypotheses enhances the validity or utility of the theory.

Application of a theory

Theories are inherently abstract and are only practically useful when applied to specific problems. Therefore, to effectively apply theory, a researcher must operationalize constructs so that they can be measured and relationships tested. The operational definition of a variable, which is discussed in more detail in Chapter 3, serves as a bridge between theory and observation (Kerlinger 1986). Constructs should be operationalized very carefully – it is at this stage where theory and empirical reality are joined. Attention to the characteristics of the population and its environment as well as particulars of the research problem is very important. For example, the use of culturally relevant survey instruments is critical at this stage. It is at this stage – the application of theory – that theoretical constructs are validly translated into variables (Glanz et al. 1997). "A key link in the struggle to have the theoretical statements and the empirical work deal with the same phenomena is sound definition of the variables" (Simon 1978).

Often, researchers have to search for a theory to help explain an empirical phenomenon of interest. This primarily involves the conceptualization process whereby investigators must find appropriate theoretical constructs to fit the empirical problem. For example, a researcher interested in examining medication adherence would use an individual behavior model to assess factors that impact medication adherence which is an individual behavior. Interventions based on such integrated theoretical frameworks would be more effective in helping patients adhere to medication regimens (Glanz et al. 1997). A program based on the transtheoretical model (discussed later in this chapter) may be effective in increasing pharmacists' participation in the provision of medication therapy management services. The choice of a suitable theoretical framework begins with identifying the research problem and ends with selecting the framework that best applies to the research problem.

There is no predominant theory that guides research in pharmaceutical practice and policy. Instead, a multitude of theories are used from various disciplines such as business and psychology, perhaps reflecting the diversity in the field itself. In the end, however, the theoretical framework chosen should be clearly understood and related to the research problem. For example, if the researcher is investigating patients' willingness to pay for pharmacist-provided medication therapy management services and presents a framework of community acceptance of pharmacist-provided medication therapy management services, there may be a discrepancy that could lead to erroneous conclusions. The misapplication of theory diminishes the investigator's understanding of the phenomenon and compromises subsequent advances in the knowledge base in the field (Dunn and Elliott 2008). In an applied, practically oriented area such as pharmaceutical practice and policy, caution should be exercised against "artificially" fitting a research problem to a theoretical framework, because the theory and hypotheses deduced from it ultimately guide the research design, data collection strategies, data analysis, and especially the interpretation of study findings.

Common theoretical models in healthcare utilization and outcomes

Many studies in pharmaceutical practice and policy research employ theoretical frameworks to study research problems. Although the focus here is on health behavior models, there are many other types of theories and models, such as systems theory and economic decision models, that are important and useful in advancing the field of pharmaceutical practice and policy. Some well-established individual-level approaches to health behavior that have been used include the Andersen Behavioral Model, the Health Belief Model, the Theory of Planned Behavior, and the Transtheoretical Model. The following is a brief overview of each model and how each can be or, in some cases, has been applied in pharmaceutical practice and policy research.

Andersen Behavioral Model

The Andersen Behavioral Model (ABM – Figure 2.1) has been used extensively in health services research (see Andersen and Newman [1973], Andersen [2008], for full discussions of the model). The ABM was developed to better understand and predict access to care. The model purports that healthcare utilization and other health behaviors (e.g., personal health practices such as diet and exercise, process of medical care such as provider behavior) are dependent on individuals' propensity to use services (predisposing), their ability to access services (enabling), and their illness level (need) on both the individual and the contextual levels, and that healthcare utilization and other health behaviors lead to some health outcome (e.g., health status or consumer satisfaction).

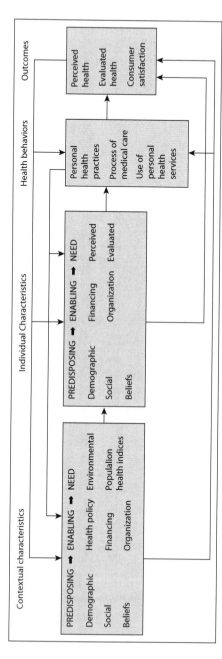

Figure 2.1 Andersen Behavioral Model. (From Andersen 2008.)

At the individual level, predisposing characteristics include demographic, social structural, and health belief variables. The enabling factors reflect individual resources and need factors to represent both perceived and evaluated need. Examples of individual level predisposing variables include age and gender, enabling variables include insurance and income, and need variables include symptoms and disease severity. At the contextual level, predisposing factors represent community-based demographic, social structural, and health belief variables. The enabling factors include the healthcare system issues of health policy and the resources and organization of the system. Need factors characterize environmental issues and population health indices. Examples of contextual level predisposing variables include community age structure, enabling variables include supply of healthcare providers, and need variables include morbidity and mortality rates. The ABM is a recursive model such that outcomes, in turn, can impact subsequent predisposing, enabling, and need characteristics and health service use (Andersen 2008).

Application of the ABM

The initial version of the ABM focused on individual level determinants and health services use. However, although still primarily a health services use model, the ABM has been expanded to include contextual level determinants, other health behaviors, and outcomes as key elements in the model. Thus, researchers who are interested in the impact of both individual and community level factors of health service use and outcomes would benefit by employing the ABM. For example, the ABM could be used to investigate the impact of pharmacists' supply and morbidity rates on medication use and incidence of drug-related outcomes. One valuable contribution of the ABM is its incorporation of both individual and contextual (community-level) elements in the decision-making process of individuals. Another advantage is that the ABM is particularly useful in guiding analyses in large database studies as well.

Health Belief Model

The Health Belief Model (HBM – Figure 2.2) consists of six dimensions: perceived susceptibility, perceived severity, perceived benefits, perceived barriers, cues to action, and modifying factors (see Janz et al. [2002] for a full discussion of the model). The most notable dimensions of the HBM, in terms of theoretical propositions and empirical investigations, are the four perception variables (Janz and Becker 1984; Harrison et al. 1992). Relationships among these perception variables are hypothesized to represent a decision process that individuals may go through when choosing among alternate health actions. The model proposes that individual beliefs about severity and susceptibility of a disease and its consequences are associated with engaging in treatment action. Subjective assessments of disease threat are assumed

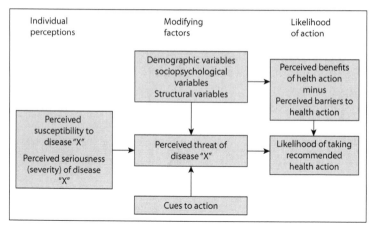

Figure 2.2 Health Belief Model. (From Glanz et al. 2002.)

to provide individuals with the motivating force to take action. Once an individual feels substantially threatened by a disease and its sequelae, alternate actions must be selected. According to the HBM, it is at this point that individuals perform a type of cost–benefit analysis such that alternatives are subjectively evaluated in terms of their benefits and costs (or barriers). This cost–benefit analysis then results in a preferred course of action (or a preferred health behavior). Self-efficacy, the confidence in one's ability to perform an action, has been added to the original HBM. Thus, one's perceived ability to overcome barriers to behavior change will also affect behavior. The two remaining components of the HBM, cues to action and modifying factors, are not conceptualized as core elements of the decision process itself, but as catalysts and shapers of the decision process. Accordingly, the HBM proposes that internal or external cues to action are needed to activate this decision-making process. Modifying factors are proposed to relate to health actions primarily through their influences on individual perceptions. Cues to action and modifying factors have seldom been examined in HBM studies and, as a result, their effects have not been well established empirically.

Application of the HBM

The HBM was originally designed to examine preventive health behaviors (e.g., vaccinations) but has since been used to investigate illness behaviors (e.g., medication adherence) as well. Contemporary issues in pharmaceutical practice and policy such as patients' willingness to participate in disease management programs and their engagement in a variety of self-management behaviors (e.g., glucose monitoring) are well-suited to studies based on the HBM. In addition, the HBM has been used in the analysis of large datasets to examine behavior. Moreover, assessments of how cultural and racial and ethnic factors impact health perceptions are included in HBM propositions.

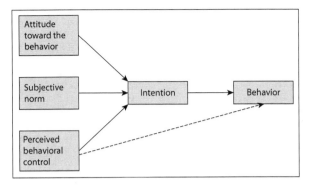

Figure 2.3   Theory of Planned Behavior. (From Ajzen 1991.)

As health perceptions are often culturally embedded and often guide people's actions about their diseases, studies based on the HBM may help to uncover key issues that could be targeted or incorporated into pharmacists' consultations or other patient-based interventions to improve health outcomes.

Theory of Planned Behavior

The Theory of Planned Behavior (TPB – Figure 2.3) is a well-established decision-making model (see Ajzen [1991] for a full discussion of the model) and is useful in predicting behaviors that are not under complete volitional control. According to the TPB, the most immediate determinant of engaging in a behavior is the intention to perform that behavior. There are three independent determinants of intentions: attitude toward the behavior, subjective norm for the behavior, and perceived behavioral control regarding the behavior. Attitudes involve the perceived advantages and disadvantages of the behavior. Subjective norms refer to how the important others or organizations feel about performing the behavior. Perceived behavioral control considers the barriers and facilitators of performing the behavior. As such, a person is more likely to engage in a behavior when he or she has a favorable attitude toward the behavior, perceives little social pressure against engaging in the behavior, and has a higher perceived control over performing the behavior.

Application of the TPB

The TPB is especially useful in investigating both cognitive and social factors that may affect behavior and has been employed across a variety of behaviors, ranging from condom use to physician-prescribing practices. The TPB gives information about the main drivers of behavior (i.e., why people hold certain attitudes, subjective norms, and perceptions of behavioral control) that can be targeted by interventions. Behaviors that primarily affect the individual are usually attitudinally driven, whereas behaviors that also impact others are more often subjective norm driven. Thus, studies employing the TPB can effectively show if behaviors of interest are primarily driven by attitudes or subjective norms and to what extent an individual has perceived control over performing the behavior. This is important because behaviors under normative control will not change with interventions targeting attitudes, and vice versa. Again, using the same example of medication adherence, the TPB could be very useful in examining cognitive and social factors related to adherence. For example, if medication adherence is found to be primarily driven by attitudes, interventions geared toward increasing adherence may need to bolster the advantages of taking medications in order to facilitate more favorable attitudes toward adhering to medication regimens.

The Transtheoretical Model

The Transtheoretical Model (TTM – constructs shown in Table 2.1) applies a stage approach to behavior change based on the perspective that behavior change is not an event but occurs through stages. It employs stages of change (also referred to as stages of readiness) to unify processes and principles of change from leading theories of behavioral change (see Prochaska et al. [1997] for a full discussion of the model). The core constructs of the TTM are stages of change, decisional balance (pros and cons), self-efficacy (confidence to change), and processes of change. According to the TTM, change is a process that involves progression through six stages: pre-contemplation, contemplation, preparation, action, maintenance, and termination.

The first stage, pre-contemplation, represents the stage where a person does not intend to change in the near future (usually within the next 6 months) and may not be aware of the need to change or have tried and failed to change behavior in the past. The second stage of contemplation represents the stage where a person intends to change in the next 6 months, but is also more aware of the pros and cons to change. At this stage, because pros and cons are equally weighed, an individual is somewhat ambivalent and tends to avoid taking action. Action-oriented programs or interventions will fail when presented to individuals in the contemplation stage. The third stage, preparation, is a stage in which people are seriously considering changing their behavior in the next month or so. People in this stage have taken steps (e.g., self-help books, plan of action) to ready themselves for behavior change and they represent good targets for recruitment into action-oriented programs. Action, the fourth stage, is the stage at which the person has changed his or her behavior and has recognized that the pros of changing outweigh the cons. The fifth stage, maintenance, is the stage at which the person has sustained the behavior for 6 months. People at this stage are less prone to relapse and are increasingly confident in their ability to maintain the change in behavior. The sixth and

Table 2.1 Transtheoretical Model constructs

Constructs	Description
Stages of change	
Pre-contemplation	Has no intention to take action within the next 6 months
Contemplation	Intends to take action within the next 6 months
Preparation	Intends to take action within the next 30 days and has taken some behavioral steps in this decision
Action	Has changed overt behavior for less than 6 months
Maintenance	Has changed overt behavior for more than 6 months
Decisional balance	
Pros	The benefits of changing
Cons	The costs of changing
Self-efficacy	
Confidence	Confidence that one can engage in the healthy behavior across different challenging situations
Temptation	Temptation to engage in the unhealthy behavior across different challenging situations
Processes of change	
Consciousness raising	Finding and learning new facts, ideas, and tips that support the healthy behavior change
Dramatic relief	Experiencing the negative emotions (fear, anxiety, worry) that go along with unhealthy behavioral risks
Self-re-evaluation	Realizing that the behavior change is an important part of one's identity as a person
Environmental re-evaluation	Realizing the negative impact of the unhealthy behavior or the positive impact of the healthy behavior on one's proximal social and physical environment
Self-liberation	Making a firm commitment to change
Helping relationships	Seeking and using social support for the healthy behavior change
Counter-conditioning	Substituting healthier alternative behaviors and cognitions for the unhealthy behavior
Reinforcement management	Increasing the rewards for the positive behavior change and decreasing the rewards for unhealthy behavior
Stimulus control	Removing reminders or cues to engage in the unhealthy behavior and adding cues or reminders to engage in the healthy behavior
Social liberation	Realizing that the social norms are changing in the direction of supporting the healthy behavior change

From Prochaska et al. (1997).

final stage is termination – a stage in which an individual is no longer prone to relapse and is completely confident that the behavior change can be sustained. This stage may be more of an ideal in most cases and is often not included in the TTM research.

According to the TTM, people progress from one stage to another through their decisional balance, their self-efficacy, and their application of different change processes. Decisional balance involves the individual's relative weighing of the pros and cons of change. Self-efficacy refers to their confidence in their ability to sustain behavior change. Processes of change are the activities or strategies that they use to help them make and maintain behavior change.

There are 10 processes that are applied through the various stages and these processes can be categorized as: cognitive and affective experiential processes, and behavioral processes. The experiential processes include consciousness raising, dramatic relief, self-regulation, environmental re-evaluation, and self-liberation. The behavioral processes are helping relationships, counter-conditioning, reinforcement management, stimulus control, and social liberation. These change processes are more or less salient at different stages and are very important to the development and content of intervention programs.

In the early stages, the cons of changing outweigh or equal the pros of changing, self-efficacy is low, and affective, cognitive, and evaluative processes (such as consciousness raising and environmental re-evaluation) are needed to progress. In the later stages, the pros of changing are higher than the cons, self-efficacy is high, and encouraging processes (such as commitments, conditioning, contingencies, environmental controls, and support) are applied.

Application of the TTM

The TTM can be employed to develop interventions aimed at promoting long-term behavior change. However, different intervention strategies are most effective at different stages of change. For example, interventions focused on contingencies (positive and negative reinforcements) would not work with individuals in the early stages of change but would be helpful in later stages (e.g., action and maintenance). The idea is to have stage matched and tailored intervention strategies so that long-term behavior change or maintenance can be achieved. The TTM has been successfully employed in smoking cessation, in the promotion of exercise, and recently in medication use studies. Another current issue in pharmaceutical practice and policy deals with enhancing pharmacists' provision of pharmaceutical care or medication therapy management services. The TTM could be useful in the development of programs that aim to increase pharmacists' provision of pharmaceutical care (Berger and Grimley 1997) or medication therapy management services. The intervention could be structured such that feedback is tailored to match pharmacists' stages of readiness (e.g., self-re-evaluation for pre-contempla-tors through techniques that promote their self-image as primary healthcare providers), and thus focused on facilitating their progress to providing the service (action) and sustaining the service (maintenance).

The ABM, HBM, TPB, and TTM are a few examples of useful models to investigate relationships of important variables and to spawn future research questions and hypotheses in the field of pharmaceutical practice and policy.

Importance of using a theoretical framework in pharmaceutical practice and policy research

Researchers utilize theories in an effort to enrich the value and interpretability of their studies. In an applied discipline such as pharmaceutical practice and policy, theories are used to inform research and practice, and findings from research and practice are then used to confirm or modify theory. Theories will facilitate the maturation of a discipline and help create a foundation on which to build. However, as noted by Mkanta and Uphold (2006), "In order to build on previous research and positively impact the state of science, it is important that researchers use a theoretical framework to guide all stages of their study."

Theories should not be "blindly" used without evaluating their adequacy in explaining pharmaceutical practice and policy phenomena (Villarruel et al. 2001). Rather, theories should be assessed for their utility in understanding phenomena in pharmacy practice situations. They must be conceptually appropriate and practically useful. Fawcett (1985) noted that "theories of a professional discipline are meaningless unless their utility as directives for practice has been established." Thus, theories should be placed in a pharma-ceutical practice and policy context, and systematically evaluated for their relevance to pharmacy practice in general and their validity to pharmacy practice problems in particular. When findings support the theory, it may be deemed appropriate for pharmaceutical practice and policy. However, if findings do not support the theory, its suitability and applicability should be questioned (Fawcett 1985).

Ultimately, the goals are to understand and, eventually, to intervene on various behaviors, be it patient or healthcare professional behavior, and to offer recommendations for healthcare practice and policy. Theories and the-oretical frameworks "explain behavior and suggest ways to achieve behavior change" (Glanz et al. 1997). In practice, pharmacy services such as medica-tion therapy management must integrate theory-based elements in order to improve drug therapy outcomes. For example, it is known from theoretically based studies that people's beliefs about medications impact their adherence to them. Thus, any medication therapy management program focusing on medication adherence should address an important variable such as patients' medication beliefs.

Types of variables in pharmaceutical practice and policy research

Definition of variable

In pharmaceutical practice and policy research, the concepts studied are called variables. Concepts (constructs) become variables in research investigations. As discussed earlier, constructs are nonobservables and variables become observables (measurable) once they are operationally defined. A variable is a concept that varies in either type or amount (Kerlinger 1986). In fact, this variation is what scientific inquiry seeks to understand – how and why things differ. On the other hand, a concept that does not vary in a research study is not a variable. For example, in studies of medication use, medication adherence is a variable because participants exhibit varying levels of medication adherence (e.g., 0–100% adherent, high versus low adherence). In the same studies, gender would be a variable because both males and females are generally included as participants. However, if similar medication adherence studies were conducted in cervical cancer patients, then gender would not be a variable because all participants would be female. In this case, gender is a constant. Other examples of variables in pharmaceutical practice and policy research include physician visits, pharmacy costs, patient satisfaction, attitudes, race/ethnicity, age, and socioeconomic status.

There are four kinds of variables that are important for research purposes. These are: (1) independent and dependent variables; (2) attribute and active variables; (3) categorical and continuous variables; and (4) latent versus observable variables. Classification of categorical and continuous variables is discussed in detail in Chapter 4. It is important for researchers to be clear about the type and classification of variables because of their importance to the conceptualization and design of the research project as well as in how the results are conveyed (Kerlinger 1986).

Independent and dependent variables

Variables play different roles in a research study and two critical roles are as independent and dependent variables. In experimental research, investigators are interested in studying the cause of some phenomenon. In an experimental design, the independent variable is the presumed cause and the dependent variable is the presumed effect. However, in nonexperimental research, the independent variable "logically" has some effect on the dependent variable (Kerlinger 1986). In other words, the independent variable does the influencing and the dependent variable is being influenced. Typical phrases such as "positively related to," "negatively correlated to," and "positively associated with" are common phrases used to indicate directional influence in nonexperimental research. Independent variables are also referred to as predictor variables and dependent variables are also called response or outcome variables.

The dependent variable is customarily the condition that researchers are trying to explain, predict, or understand. Variations in the dependent variable are presumed to depend on variations in the independent variable. Researchers, policy makers, and other stakeholders in the field are interested in the medication use process and patient outcomes. As such, a researcher might investigate the impact of antidepressant use on the incidence of weight gain. The independent variable is antidepressant use and the dependent variable is weight gain. This presumes that variations in weight gain will depend on variations in the use of antidepressants.

Variables are not inherently independent or dependent. They can be independent in one study and dependent in another study, and even change within the same study. For example, a study might investigate the relationship between gender (independent variable) and medication adherence (dependent variable). Another study might examine the effect of medication adherence (independent variable) on high blood pressure (dependent variable). Therefore, the classification of independent or dependent variables "is more a classification of *uses* rather than a distinction between different kinds of variables" (Kerlinger 1986).

Attribute and active variables

Independent variables are further classified as attribute or active. Many variables studied in pharmaceutical practice and policy research are attribute variables. Attribute variables are those characteristics that participants bring to a research study and are simply measured or documented by the researcher. Attributes of a person (e.g., race, socioeconomic status, gender) or environment (e.g., rural/urban, county or region of residence) are examples of attribute-independent variables. Attribute variables are typically not actively manipulated in the research.

Active variables, on the other hand, can be manipulated by the researcher. Active variables are most relevant in experimental research whereby study groups are exposed to different treatments, the manipulated variable. For example, in a study of the effects of a pharmacist-provided medication therapy management service, one group is exposed to the service (treatment group) and one group does not receive the service (control group). In this case, pharmacist-provided medication therapy management service is the active independent variable.

Although some variables are naturally attributes, other variables can be either attribute or active depending upon the study situation (Kerlinger 1986). Attitude is a good example. A person's attitude toward taking medications can be measured (attribute) or actively manipulated as in the following example. A researcher may promote a more favorable attitude toward a medication by telling members of one group that the drug is highly effective with few side effects whereas members of the other group are told that the drug is highly effective but has many side effects, promoting a less favorable attitude. Other examples of variables that can be attribute or active include blood pressure and anxiety levels. The distinction between attribute and active independent variables is an important one because it is critical to the research design and implementation (Kerlinger 1986) and to the evaluation of cause-and-effect relationships in a research investigation (Harmon and Morgan 1999).

Independent and dependent variables can be continuous or categorical. Although a part of the conceptualization phase, this distinction is most relevant to data analyses and thus is discussed in more detail in Chapter 11. Suffice it to say, the appropriateness of the data analysis procedure is dependent on whether the dependent variable is continuous or categorical.

Latent versus observable variables

A final classification of variables involves latent versus observable variables. As implied in the earlier discussion of theory, "a latent variable is an unobserved 'entity' presumed to underlie observed variables" (Kerlinger 1986). The descriptions of theories and hypotheses are referring to relationships of latent variables. Latent variables can be thought of as constructs. Researchers interested in patient satisfaction must understand that patient satisfaction is a latent variable. One cannot see it or feel it. In order to study patient satisfaction, however, it must be measured. In a research study, the presumed indicators of patient satisfaction are actually measured, such as a score on a satisfaction scale. Patient satisfaction itself cannot directly be measured because, as Kerlinger (1986) noted, it is an in-the-head variable, an unobservable thing – hence, a latent variable. It is not until patient satisfaction is operationally defined that it becomes an observable variable.

Scientific relationships in pharmaceutical practice and policy research

Researchers are generally interested in how one or more phenomena relate to other phenomena. Namely, they are interested in exploring and testing relationships (Polit and Beck 2008). A relationship exists between two variables when a change in one variable corresponds with a change in the other. The association can be positive or negative, e.g., it is fairly well established that there is a negative relationship between health insurance coverage and morbidity rates. More specifically, a lack of health insurance coverage (independent variable) is related to increased morbidity (dependent variable). When conducting quantitative research studies, investigators are testing the relationships between independent and dependent variables.

Correlational and causal relationships

The nature of these relationships can be correlational or causal. A correlational relationship is one in which changes in the independent variable are associated with changes in the dependent variable. Using the example above, although one cannot necessarily say that a lack of health insurance causes morbidity, one can say that the two variables are correlated with each other. On the other hand, a causal (or cause-and-effect) relationship is one in which the independent variable causes the dependent variable or, to put it another way, a change in the independent variable produces a change in the dependent variable. A simple example is that increased caloric intake causes weight gain. It should be noted, however, that, although true causality cannot be established (Kerlinger 1986), there are methods that researchers can employ to empirically validate their causal notions (see Chapter 5 for more detail).

Connecting theory and variables in pharmaceutical practice and policy

In the area of pharmaceutical practice and policy, often research problems and important variables relevant to the research problem are initially identified. Then, an appropriate theoretical framework that could elucidate the problem and variable relationships is sought. As previously mentioned, the key to selecting a theoretical framework is connecting variables with appropriate constructs. Constructs are conceptually defined as part of theoretical frameworks and these definitions reveal the meaning of the constructs in the context of the respective framework. Given these definitions, variables to be investigated should be logical empirical indicators of the constructs as they are defined in the theoretical framework or a variable should be a "suitable representation" of the construct for which it stands (Hox 1997). For example, a researcher is interested in studying patient decision-making relative to medication adherence – an empirical indicator of illness behavior. That researcher should consider individual decision-making frameworks such as the TPB that are used to examine many different types of behaviors, including health and illness behaviors. Another investigator who wants to examine how factors such as health insurance (an empirical indicator of individual factors) and urban environment (an empirical indicator of contextual factors) impact individuals' use of pharmacy services (an empirical indicator of health service utilization) should consider using ABM to better understand how individual and contextual factors influence health service utilization, the hallmark of the ABM.

Connecting theory and variables appropriately is dependent on the researcher's understanding of the conceptual underpinnings of the model as well as the practical context of the research problem. However, there are various other resources that can facilitate the researcher in making the connection between the theory and problems in a research area. Literature reviews on the topic area often provide frameworks that are appropriate to the study. In addition, experts in a particular field are a good source of information about appropriate and suitable theoretical frameworks to study research problems in that field. Generally, it should be noted that multiple frameworks may be appropriate to study a research problem. In fact, studying a research problem from more than one conceptual viewpoint can lead to a broader understanding of the problem.

Summary and conclusions

When conceptualizing research, the use of theory is a critical and necessary step. Theory tells a story and provides a blueprint for research investigations. Establishing a foundation of knowledge based on systematic and integrated science is essential. Throughout this chapter, the importance of using theoretical frameworks to the advancement of pharmaceutical practice and policy research has been emphasized. This chapter also described several theoretical models that can be applied in pharmaceutical practice and policy research. Theoretically inspired research using models such as the ABM and TPB may provide different perspectives and uncover relationships that are critical to practice and policy (Mkanta and Uphold 2006). However, care must be taken to correctly apply theory to research problems in a practically oriented field such as pharmaceutical practice and policy. The practical and theoretical value of findings of scientific investigations is directly related to researchers making appropriate connections between theory and variables such that advancements in knowledge are not compromised. Research shapes practice and policy, which in turn inform research and conceptualization of future studies. "The gift of theory is that it provides the conceptual underpinnings to well-crafted research and informed practice" (Glanz et al. 2002). The growth of the discipline and its impact on practice and policy will be heavily influenced by the consistent use of theory and rigorous evaluation of its utility, with a focus on making theoretical refinements as well as practice and policy modifications. Such activities will advance pharmaceutical practice and policy in the effort to positively impact public health in an evolving healthcare environment.

Review questions/topics

1 What is theory?

2 Describe an example of a theoretical model presented in this chapter and discuss how it might be applied in pharmaceutical practice and policy research.

3 Why is it important to use theoretical models in pharmaceutical practice and policy research?

4 Describe the different types of variables and give examples of each.

5 Discuss how a researcher might evaluate the appropriateness of a theoretical model to study a research problem.

References

Ajzen I (1991). The theory of planned behavior. *Organizational Behavior and Human Decision Processes* 50: 179–211.

Andersen RM (2008). National health surveys and the behavioral model of health services use. *Med Care* 46: 647–53.

Andersen RM, Newman J (1973). Societal and individual determinants of medical care utilization in the United States. *Milbank Mem Fund Q* 51: 95–124.

Berger BA, Grimley D (1997). Pharmacists' readiness for rendering pharmaceutical care. *J Am Pharm Assoc* 37: 535–42.

Brazil K, Ozer E, Cloutier MM, *et al.* (2005). From theory to practice: improving the impact of health services research. *BMC Health Serv Res* 5: 1.

Crosby RA, Kegler MC, DiClemente RJ (2002). Understanding and applying theory in health promotion practice and research. In: DiClemente RJ, Crosby RA, Kegler MC (eds), *Emerging Theories in Health Promotion Practice and Research*. San Francisco, CA: Jossey-Bass, 1–15.

Dunn DS, Elliott TR (2008). The place and promise of theory in rehabilitation psychology research. *Rehabil Psychol* 53: 254–67.

Fawcett J (1985). Theory: basis for the study and practice of nursing education. *J Nurs Res* 24: 226–9.

Glanz K, Lewis FM, Rimer BK, eds (1997). Theory, research, and practice in health behavior and health education. In: *Health Behavior and Health Education: Theory, research, and practice*, 2nd edn. San Francisco, CA: Jossey-Bass, 22–39.

Glanz K, Champion VL, Stretcher VJ (2002). The health belief model. In: Glanz K, Rimer BK, Lewis FM (eds), *Health Behavior and Health Education: Theory, research, and practice*, 3rd edn. San Francisco, CA: Jossey-Bass, 45–66.

Harmon RJ, Morgan GA (1999). Research problems and variables. *J Am Acad Child Adolesc Psychiatry* 38: 784–5.

Harrison JA, Mullen PD, Green LW (1992). A meta-analysis of studies of the health belief model with adults. *Hlth Educ Res* 7: 107–16.

Hox JJ (1997). From theoretical concept to survey question. In: Lyberg LE, Biemer P, Collins M, *et al.* (eds), *Survey Measurement and Process Quality*. Hoboken, NJ: John Wiley & Sons, Inc., 47–69.

Janz NK, Becker MH (1984). The health belief model: a decade later. *Hlth Educ Q* 11: 1–47.

Janz NK, Champion VL, Strecher VJ (2002). The health belief model. In: Glanz K, Rimer BK, Lewis FM (eds), *Health Behavior and Health Education: Theory, research and practice*, 3rd edn. California: Jossey-Bass, 45–66.

Kerlinger FN (1986). *Foundations of Behavioral Research*, 3rd edn. New York: Holt, Rinehart & Winston.

Mkanta WN, Uphold CR (2006). Theoretical and methodological issues in conducting research related to healthcare utilization among individuals with HIV infection. *AIDS Patient Care STDS* 20: 293–303.

Polit D, Beck C (2008). *Nursing Research: Generating and assessing evidence for nursing practice*, 8th edn. Philadelphia: Lippincott Williams & Wilkins.

Prochaska JO, Redding CA, Evers KE (1997). The transtheoretical model and stages of change. In: Glanz K, Lewis FM, Rimer BK (eds), *Health Behavior and Health Education: Theory, research, and practice*, 2nd edn. San Francisco, CA: Jossey-Bass, 99–120.

Simon JL (1978). *Basic Research Methods in Social Science*, 2nd edn. New York: Random House.

Villarruel AM, Bishop TL, Simpson EM (2001). Borrowed theories, shared theories, and the advancement of nursing knowledge. *Nurs Sci Q* 14: 158–63.

Online resources

National Cancer Institute. *Theory at a Glance: A guide for health promotion practice*, 2nd edn. available at: www.cancer.gov/PDF/481f5d53-63df-41bc-bfaf-5aa48ee1da4d/TAAG3.pdf.

3

Operationalizing research

Khalid M Kamal

Chapter objectives

- To identify the characteristics of a research problem
- To identify the sources of pharmaceutical research problems
- To describe the process of identifying good research questions
- To identify the characteristics of a scientific hypothesis
- To compare and contrast types of hypotheses
- To operationalize variables and concepts

Introduction

Healthcare research is a systematic process that comprises an ordered set of activities and is focused on addressing problems and issues that have implications for the patient, caregivers, and society. Researchers are interested in understanding cost, access, and quality of pharmaceutical care from diverse perspectives by asking questions and searching for answers. The ability to ask creative questions and generate new knowledge for pharmaceutical practice and policy involves a sequence of steps starting from conceptualizing a research project to implementing and disseminating the findings.

Broadly speaking, research comprises two phases: the thinking phase and the action phase (DePoy and Gitlin 1998). The focus of the thinking phase is on identifying a problem, formulating a research question and hypothesis, and identifying pertinent study variables. This step is important because it provides direction to the action phase of research, which mainly deals with the study methodology, analyses, and dissemination of findings. The research process can be considered as being circular in nature rather than linear (Fain 2004). There is a lot of back-and-forth movement involved between the different activities in both the phases. In the thinking phase, researchers continually evaluate ideas, review literature, rethink, and reconceptualize problems before finalizing the study problem statement. In the action phase, study results help to create a new knowledge base or generate new ideas for the next study.

The purpose of the chapter is to describe how to formulate and articulate the thinking phase of research. This chapter focuses on the importance of research problems and discusses some sources that can be used to help formulate a good research problem. Next, definition and development of research questions and hypotheses are discussed. The chapter concludes with a discussion and direction for conceptually and operationally defining variables in a research study.

Steps in operationalization

The first and most significant step in the thinking phase of a research process is identifying and selecting a research problem. Research problems provide the purpose of the study and give direction to the subsequent steps of the research process. Having identified a problem that is clear and researchable, the next step is to conduct a thorough review of the literature, the major purpose of which is to identify what is and is not known about the problem area. It also helps researchers to determine how a problem can be studied.

After a research problem has been selected and further refined, based on a literature review, the next two steps are to formulate research questions and hypotheses. These also identify the variables and the relationship of the study variables. To move to the final step, a researcher has to identify and create conceptual and operational definitions of the concepts and variables involved in the study. Operational definition is a clear and concise definition of a measure and is fundamental to the data collection stage. Each of the four steps, namely research problem, research question, hypothesis, and operational definition, is discussed in detail below.

Research problems

Selection and definition of a research problem

A problem is defined as an interrogative statement that questions the interrelationship of two or more variables. Other than experimental psychology (e.g., memory), most research problems have two or more variables (e.g., adherence, pain management). Well-defined problems do not appear spontaneously and it may take researchers months or even years of exploration and thought to come up with a complete problem that can be researched.

At this point, it is important to understand the difference between a research topic and a research problem. Most researchers starting out easily identify a broad area of interest or a research topic, examples of which include screening services in ethnically diverse communities, pain management, and outcomes research such as quality of care, health status, and impact of new technologies on costs. As seen from these examples, a single research topic can yield a number of research problems. It is important to note that research problems shape the development of each subsequent step of the research process and, thus, must be carefully formulated. According to Norwood (2000), a well-written research problem statement should be able to answer the following three questions:

1   What is being studied?
2   Who is being studied?
3   Why is this being studied?

The first question is designed to identify a research topic that is to be studied. A researcher starts with a general topic of interest or problem to be solved and continuously refines it to have a more focused topic. For example, a broad area of interest could be stated as "studying influenza immunization rates." Although this is a good start, it lacks focus and more information is needed before a research problem can be formulated.

The second question identifies the population in which the study would be conducted. The goal is to characterize the population in terms of social, demographic, and economic characteristics. This helps a researcher focus on issues that are prevalent in a given population segment or community. To further develop the research problem, it can be restated as "studying influenza immunization rates among inner-city poor populations."

The third question emphasizes the relevance or significance of the study and its contribution to the knowledge base. To elaborate on the immunization example, it might already be known that influenza immunization rates are low among inner-city poor populations. The issue then is to study the reasons for the poor uptake of immunization in this population. This new knowledge can significantly help in designing new interventions or bring about policy changes so as to improve the immunization rates. The final research problem can be "Why are immunization rates among inner-city poor populations the lowest in the country?"

Once one or more potential problems have been identified, they undergo several iterations until they are ready for empirical investigations. These iterations are important in refining and narrowing the focus of potential research problems and improving their significance. The process of refining and narrowing the original problem requires a lot of thought and action, both of which can occur simultaneously (Stamler 2002). To illustrate the thinking and action process, consider the following sample research problem: "What is the impact of physician counseling on smoking cessation attempts after discharge from the hospital?" The broader research topic may well have been to identify strategies to improve quit rates after hospital discharge. Thought and action processes can be used to extract a problem from the complex research topic that focuses on the impact of physician counseling.

The action process of refining the research problems includes several strategies (Norwood 2000; Stamler 2002). A research problem has the potential to generate a number of studies depending on the researcher's perspective or interests. Thus, asking probing questions will help clarify the problem and provide direction to the research study. Another strategy is to critically review the literature, identify gaps in knowledge, and refine the research problem. Literature review also provides sources to justify the research strategy. Finally, discussing one's ideas with mentors, experts, or peers can help generate new ideas or uncover pitfalls in a research study. During all these steps, one may move between the thought and action process several times while continuously refining the problem (Stamler 2002).

Characteristics of a good problem

According to Kerlinger and Lee (2000), a good problem has the following three characteristics: first, a problem should express a relationship between two or more variables; second, the problem should be well defined, clear, and free of any ambiguity; and, third, the problem must be ready for empirical investigation. This means that the relationship between variables should be stated and the variables should be measurable. Selecting a research problem is the first step in the research process and is also the most difficult step. It takes a lot of time, thought, and effort to identify a research topic and focus on a researchable problem. This overwhelming task can be made easier, if one knows where to find these research problems.

Sources of research problem

A variety of sources can be used to help formulate a research problem. These include use of theory, clinical experiences, prior research, case studies, conflicting findings, and replication. A study by Stamler (2002) reports the use of 12 possible sources of research problems in the nursing literature, originally identified by Gillis and Jackson (2002). The sources include:

1 improving clinical practice
2 exploring a pattern of incidents
3 testing folk wisdom
4 understanding phenomena from the insider's perspective
5 tackling current issues
6 inconsistencies in the literature
7 testing a theory
8 testing practice theories
9 exploring variations in a dependent variable
10 providing an evaluation
11 implementing and studying action
12 replicating a study.

Another useful source for research ideas is funding agencies such as the government, foundations, or pharmaceutical companies. These funding agencies always invite study proposals and interested researchers can design a study to fit within the expectations of the funding agency.

Research questions

A research question is an explicit inquiry that yields hard facts to help challenge, examine, analyze, or solve a problem and produce new research (Beitz 2006). Research questions are interrogative statements and, specifically, indicate what researchers want to find out about a problem area. Research questions focus on describing variable(s), examining relationships among variables, or determining differences between two or more groups on a single or multiple variable(s) of interest (Fain 2004). Some examples of research questions include:

1   What is the incidence of influenza in white and nonwhite people?
2   What are the differences in health beliefs about influenza vaccination immunization between white and nonwhite people?

Characteristics of a well-written research question

In contrast to research problems, research questions are very specific and precise. Research questions should be well written and clinically relevant, and have the potential to yield information to improve pharmaceutical practice and policy. The characteristics of a well-written research question can be summarized by the mnemonic FINER which stands for feasible, interesting, novel, ethical, and relevant (Kwiatkowski and Silverman 1998; Beitz 2006; Cummings 2007).

Feasible

There are important issues that must be considered, and are often overlooked, before starting a research project. The success of the project depends on feasibility issues such as enrollment, multidisciplinary approach, technical expertise, time requirements, and funding. Enrollment of participants is an important issue that can help achieve a study's intended purpose. Another important factor is the researchers' technical skills and experiences which are required right from project conceptualization to dissemination of the study results. Adequate time and resources such as money, staff, and supplies must be estimated accurately to ensure that the participants are available, and the project is completed ethically and within budget. Finally, researchers need to stay focused on the project goals and not try to answer too many questions or collect data that were not originally intended.

Interesting

Research is time-consuming and requires substantial energy and motivation. There could be delays involved, which could result in frustrations with the process. To produce high-quality systematic research, researchers should be passionate about their work and should maintain a good intensity of interest and focus throughout the research process.

Novel

Any new research study must have the potential to add to the existing knowledge base in pharmaceutical practice and policy. Occasionally, studies can be replicated to see if the results are generalizable to a new setting or a different population group, or if a new methodology builds on an existing technique. To determine if the study is novel, a researcher should conduct an exhaustive literature review, discuss with experts in the field, or search for studies that have been supported by funding agencies.

Ethical

A good research question must meet ethical standards. In the USA, the Institutional Review Board must review a research study for safety, privacy, confidentiality, and consent before approving it. This review process is important in protecting the rights of the participants, reducing their risks, and increasing their benefits, both in carrying out the research and in publication of the results. Special care must be given to studies using vulnerable groups, such as prisoners, people with learning disabilities, other cultures, to ensure ethical treatment.

Relevant

As mentioned earlier, the study has to add to the scientific knowledge base of pharmaceutical practice, influence pharmaceutical policies, or advance further research work. If the study is unable to produce such results, it is irrelevant and should not be pursued. If faced with studies of uncertain relevance, it is always advisable to discuss them further with mentors, clinicians, or experts. The questions should be very specific and measurable.

Process of framing research questions

The process of framing a specific yet comprehensive research question can overwhelm any new researcher. This process can be made easy if researchers refrain from using highly complex questions as their first effort. Complex research questions need complicated research design, which can make any research process look difficult. Instead, researchers should eliminate all irrelevant distractions, no matter how interesting, from the broad topic and select a simple and clearly stated question.

Before a simple research question is stated, researchers should know how to frame a question. There are no standard guidelines for writing questions; however, as the questions have a profound effect on the research process, it is advisable to be familiar with the basic rules of writing questions that are clear and concise. There are two components to any research question: the stem of the question, which directs the research process, and the topic of interest, which is simply what the question is about or the focus of the study (Brink and Wood 2001). For example, in the question "What are the characteristics of successful practitioners?" the stem is "what are" and the topic of interest is "characteristics of successful practitioners." Based on the number of stems and topics, a research question can be classified as a simple or complex question. The above sample question is a simple question because it has one stem and one topic. A complex question, on the other hand, has more than one stem and topic. For example, "What is the relationship between exercise and obesity?" is a complex question because it has two topics of interest – exercise and obesity.

Apart from being simple, a research question must be action oriented, i.e., a question must demand an answer. An example of an action-oriented question is "What is the relationship between exercise and obesity?" The same question can be presented as a statement "Exercise has an effect on obesity." However, the statement is simply stating a fact and requires no further research. Instead, if the statement is presented as an action-oriented question, it demands an answer. There are certain questions that help little to answer a research topic and must be avoided. These questions are called "stoppers." A question that can be answered by a "yes" or "no" is not considered action oriented because there is no need to do any research to find the answer. Similarly, questions that start with "should," "could," or "do" are also not considered action oriented because they elicit opinions and not facts (Wood and Ross-Kerr 2006). For example, "Should obese patients exercise?" or "Do obese patients exercise?" can be answered as a "yes" or a "no" and are not action oriented. However, by using stems such as "what" or "why," the question (stopper) can be changed to an action-oriented question by changing from a passive to an active voice. The other half of the research question, i.e., the topic of interest, can be simple or complex depending on whether there is one concept or multiple concepts or ideas.

Types of research question

At this point, it is clear that identifying the right question is one of the most important steps in a research process. Working with the right question is also important because the type of question dictates the use of specific research approaches (Meadows 2003; Newton et al. 2004). These approaches can be broadly categorized into quantitative and qualitative.

Quantitative research is a systematic process that utilizes numerical data to obtain information about the problem and is often theory driven. Randomized clinical trials and surveys are examples of quantitative designs. On the other hand, qualitative research involves descriptions of cases or social phenomena in a natural setting and do not use calibrated or scaled measurements as quantitative research does. For example, differences in the immunization rates in elderly populations across ethnic groups can be examined using quantitative research such as a survey. If the researchers, instead, are interested in identifying factors that influence the disparity in immunization rates in ethnically diverse population, a qualitative research such as an in-depth interview can be used. Research generally addresses three basic types of research questions: descriptive, relational, or causal (Meadows 2003).

Descriptive

The primary purpose of a descriptive study is to describe what is going on or what exists. Such a study forms the basis for all research. Examples of descriptive studies include case studies and prevalence and incidence studies. For example, a study examining the prevalence of smoking among general practice patients and assessing their readiness to quit is a type of descriptive study. Descriptive questions can be answered by both qualitative and quantitative approaches.

Relational

A relational study is designed to look at the relationship between different phenomena. For example, the proportion of men willing to quit smoking compared with women in the last year is essentially studying the relationship between gender and willingness to quit smoking. Relational questions can be answered only by quantitative research.

Causal

This type of study examines whether one or more variables (e.g., a program or treatment variable) cause or affect one or more outcome variables. The goal of causal studies is to determine cause and effect. For example, does an intervention program using a motivational interviewing technique improve quitting in a smoking population? In such a study type, it is important to ensure that the only possible cause is the variable of interest. Causal questions can be answered only by quantitative research. Randomized controlled trials are considered the gold standard technique for establishing cause-and-effect relationships.

Hypothesis

Definition

According to Kerlinger and Lee (2000), a hypothesis is a specific statement of prediction presented in a testable form that consists of at least one specific relationship between two variables (concepts), which are measurable. For example, "Patients with a greater knowledge of asthma have a significantly higher rate of adherence to the medication regimen." The hypothesis identifies disease knowledge and adherence as two variables with a relationship that can be observed and measured.

When there is enough information to predict the outcome of the study and the researcher intends to test the significance of the prediction, the research question is stated as a hypothesis. A clear and concise question can be converted into a hypothesis by using a testable, declarative statement which can be either supported or refuted (Fain 2004; Lipowski 2008). By specifying a prediction about outcome, a hypothesis connects theory (abstract) to the empirical world (concrete) (Polit and Hungler 1991). The process of generating a hypothesis thus involves moving from a generalized set of theoretical statements to establishing a more concrete and specific relationship between variables to generate an answer (Behi and Nolan 1995). If the observed relationship between variables is consistent with the study hypothesis, then there is empirical evidence in support of the theory. On the other hand, if the observed relationship between variables is not consistent with those hypothesized, it is an indication that the theory needs to be further refined and verified.

Hypotheses are mostly associated with quantitative research where they are generally derived from a theory and presented as a specific and precise statement. Hypotheses testing also occurs in qualitative studies but, compared with quantitative research, hypotheses differ in nature and form. In qualitative studies such as exploratory or descriptive studies, hypotheses are derived from literature review or observations and are mainly used for theory building (Behi and Nolan 1995). Sometimes, qualitative studies are considered as studies with no hypothesis. On the other hand, there can be studies with more than one hypothesis.

Regardless of the format used to state a hypothesis, the declarative statement should include the variables being studied, the population being studied, and the predicted outcomes. According to Fain (2004), a general model for stating hypotheses is as follows: study participants who receive X are more likely to have Y than those who do not receive X. In this model, X represents the independent variable (variable manipulated by the researcher) and Y is the observed outcome. Words such as "greater than," "less than," "positively," or "negatively," which are commonly used in a hypothesis, simply denote the direction of the hypothesis. For example, patients with greater knowledge of asthma have a significantly higher rate of adherence to the medication regimen than patients who have little knowledge about their disease:

Study participants = patients with asthma
X   = knowledge of the disease
Y   = adherence rate.

Criteria for a good hypothesis

There are several criteria that must be considered in order to formulate a good hypothesis (Behi and Nolan 1995). First and foremost, the hypothesis must be written *a priori* – meaning before the research. As hypothesis exerts considerable influence on different aspects of the study, an *a priori* hypothesis will help researchers maintain objectivity throughout their research process. Second, the hypothesis must be stated very clearly, with clear operational definitions of variables. Third, a single hypothesis should specify only one relationship. As a hypothesis with multiple relationships among variables is difficult to understand, the researcher can use multiple hypotheses, each expressing a single relationship between two variables. This will help increase the clarity about the relationship between multiple variables and it will also be easy to confirm or refute study findings. Fourth, the hypothesis must be testable and presented in a declarative format, so that a research study can be designed to test the predicted relationship between variables. Finally, the hypothesis should state the relationship between the study variables. Generally, the relationship is causal in nature; however, if an associative hypothesis is suggested, the positive or negative correlation has to be stated as well.

Types of hypotheses

Hypotheses are classified as simple versus complex, nondirectional versus directional, associative versus causal, and statistical versus research (Norwood 2000; Fain 2004).

Simple versus complex hypotheses

A simple hypothesis predicts the relationship between one independent variable and one dependent variable. A complex hypothesis predicts the relationship between two or more related independent variables and/or two or more related dependent variables. Although simple hypotheses are easier to test, measure, and analyze, they are less useful in our research field given the range of factors that are considered in a research study. A researcher has to decide the type of hypothesis that is best for the study and for the study's feasibility. If a complex hypothesis is appropriate but it is difficult to test or collect data, then a simple hypothesis should be considered instead. For example, a simple hypothesis can be stated as "Patients with greater knowledge of asthma have a significantly higher rate of adherence to the medication regimen than patients who have little knowledge about their disease." The independent variable is the amount of knowledge (greater versus lesser) and the dependent variable is adherence. A complex hypothesis is stated as "Patients with greater knowledge of asthma and better access to care have a significantly higher rate of adherence to the medication regimen, decreased hospitalization rate, and incur lower costs of care than patients who have little knowledge about their disease or have a barrier to access to care." In this hypothesis, the independent variables are the amount of knowledge (greater versus lesser) and access to care (some versus none), and the dependent variables are adherence, incidence of hospitalization, and costs of care.

Nondirectional versus directional hypotheses

In a nondirectional hypothesis, a relationship between variables is predicted but the researcher is unsure or not confident about the direction of the outcome. It is difficult to identify independent and dependent variables in such hypotheses. Nondirectional hypotheses are used when past research provides inconsistent or conflicting results or when the direction of the relationship is unknown and is simply stated as "X is related to Y." In a directional hypothesis, the direction of the relationship between independent and dependent variables is clearly stated.

A directional hypothesis is stated as "When X increases, Y increases" (positive or direct relationship) or "When X increases, Y decreases" (negative or indirect relationship). Thus, these hypotheses are more logical compared with nondirectional hypotheses. For example, if social support is related to depressive symptoms in elderly people, this is considered a nondirectional hypothesis. It does not indicate whether the depressive symptoms increase or decrease with the presence or absence of social support for elderly people. On the other hand, if it states that the presence of social support decreases depressive symptoms in elderly people, then this is a directional hypothesis.

Associative versus causal hypotheses

The basic difference between associative and causal hypotheses is whether a cause and an effect have been specified (Behi and Nolan 1995). In an associative hypothesis, there is no implication for causality and it simply asserts that a change in one variable is associated with change in the other (Norwood 2000). This association can be either positive or negative. In positive associations, variables move in the same direction whereas, in negative associations, variables move in the opposite directions. For example, research shows that patients with chronic conditions such as arthritis or diabetes have depression. Thus, there seems to be a positive correlation between chronic conditions and the presence of depression.

In a causal hypothesis, both independent and dependent variables are used to describe the nature of the predicted relationship. Causal hypotheses are always directional and independent variables must precede dependent variables. In addition, causal hypotheses are stated as null or alternate hypotheses. For example, a causal hypothesis can be stated as "A new treatment regimen will cause a difference in depression." This describes the cause-and-effect relationship between the treatment regimen (independent variable) and depression (dependent variable).

Statistical versus research

A statistical hypothesis, also referred to as a null hypothesis, asserts that no relationship exists between two variables. Statistical hypotheses are usually used to test the relationship statistically and determine whether the observed relationship is due to chance or is a true relationship. A research hypothesis, also referred to as an alternate hypothesis, asserts that a relationship or difference exists between two variables. This type of hypothesis indicates what the researcher expects to find in the study results. Research hypothesis can be simple or complex, directional or nondirectional. The example used for a causal hypothesis can be stated as a null hypothesis "There will be no change in depression as a result of the treatment regimen." The alternate hypothesis will be stated as: "There will be a change in depression as a result of the treatment regimen."

Operationalization

Abstraction and extensionalization

The research problems, research questions, and hypotheses all identify variables or concepts that can be measured or manipulated within the study (Burns and Grove 2005). Abstraction is the process of reducing the information content of variables and concepts deemed unimportant to a clear definition. The process of abstraction involves some loss of detail and, therefore, the derived concept is never equivalent to the original concept (Hepler 1980). A variable is merely a quality or a characteristic of a person, thing, or situation that can be measured, e.g., height, weight, and blood pressure. Variables are also concepts at various levels of abstraction and are classified into two types: highly concrete or lower-level abstractions that have distinct characteristics and are easily measurable (e.g., blood pressure, height) and highly abstract or higher-level abstractions (e.g., love, anger, panic). The abstract concepts, even within a discipline, have several meanings, are vague, and need further refinement before they can be measured. The higher-level abstractions are referred to as research concepts and are generally studied in qualitative studies and some quantitative studies (correlational). The observable characteristics of a concept are called indicators. A given observable characteristic can also be an indicator for several concepts, e.g., chest pain can be an indicator for angina, indigestion, ulceration, or musculoskeletal injury.

When a hypothesis is being developed, the process of abstraction is continuously occurring and the hypothesis is linked by the highest levels of two or more abstractions. For example, "compliance" and "counseling" in the following hypothesis are the highest levels of abstractions: "Patients receiving counseling from their pharmacists are more compliant with medications compared with patients who are not counseled by pharmacists." The two higher levels of abstractions have to be refined and converted to lower-level abstractions so that they can be observed and measured in the study. For example, compliance can be measured from patient self-reports, pill counts, urine tests, etc. Similarly, counseling can be measured as oral or written communication, formal or informal styles, etc. The process of keeping the abstractions distinct and maintaining the relationship between higher level abstractions and observable facts is called extensionalization, i.e., extensionalization is the connection between reality and the perception of the observer. Mixing two levels of abstractions can be misleading and should be avoided (Hepler 1980).

In the above example, each of these abstractions is measured in a number of ways. As each of these measurements will yield a different study result, it is important to extensionalize the abstractions. One method of extensionalization is by ostention, the definition of which is accomplished by showing or pointing to it as it is named (Hepler 1980). This is used in the early stages of extensionalization, and is often used in radiological, microbiological, or dermatology studies using pictures and figures. Even though this procedure seems archaic, it is a useful method to relate labels to reality (Hepler 1980). Another important method of extensionalization is to use an operational definition. Here, an abstraction is defined by the procedures or actions necessary to quantify or measure the variables and concepts. An operational definition develops a concrete and procedural definition of abstract concepts and variables that otherwise are difficult to measure. This is often used in pharmaceutical practice and policy research.

The process of operationalizing a variable or a concept requires the development of conceptual or operational definitions, based on theoretical meaning, into a specific concept or variable. A conceptual definition is more comprehensive than a denotative definition and is established through concept synthesis or concept analysis (Burns and Grove 2005). Intelligence, honesty, weight, and knowledge are examples of ideas that are often conceptually defined. Once a conceptual definition has been established, the researcher must then use an operational definition to indicate how the abstract concept will be measured. For example, one of the conceptual definitions of weight would be a unit measure of gravitational force. This definition provides a basis for formulating an operational definition such as measurement of an object on a weighing scale in kilograms or pounds.

Operationally defining a variable compels the researcher to express abstract concepts in concrete terms. The selection of measures to determine the concepts will depend on the degree of abstraction. It is easier to operationalize a concrete concept such as "height" than an abstract concept such as "anger." Height can be operationally defined in centimeters or inches but there is less clarity in how anger can be measured. Anger can be defined operationally in facial expression or tone of voice, but the definition will differ across people, making it difficult to operationalize. Generally, abstract concepts such as anger or satisfaction are measured by indices where several discrete indicators are combined into one score (using standardized questionnaires or scales).

Operational definition is usually derived from a set of procedures or activities employed by the researcher in either manipulating an independent variable or measuring a dependent variable (Kerlinger and Lee 2000; Burns and Grove 2005). Based on whether the variables in the study are manipulated to create the presence of the independent variable or describe the measuring operations for the dependent variable, operational definitions can be classified into experimental and measured, respectively. For example, consider the hypothesis "Patients are more compliant with medications if the group counseling session is short, rather than long." The behavior (dependant variable) "compliant" and the duration of session (independent variable) "short" or "long" need to be operationalized. The experimental operational definition will spell out the details of the researcher's manipulation of the independent variable. Intervention researchers often spend significant time and effort in developing the procedures and details of experimental operational definitions (short session of counseling [including details] for 30 minutes or less, and long session of counseling [including details] for more than 30 minutes). The measured operational definition will describe the operations by which the dependent variable is measured. The measured operational definitions are often driven by the data collection and coding procedures (compliance expressed as medication possession ratio calculated as the number of days of medication dispensed as a percentage of 365 days). Nonexperimental research involves measured operational definitions for all variables.

Issues to consider in operationalization

There are certain issues that need to be considered when operationalizing a research question or hypothesis. Most of the time, a study will include concrete concepts that can be easily operationalized. Concepts in quantitative research are very specific and can be quantified (converted to numbers) with ease. The concepts in qualitative research are usually more abstract in nature and are broadly defined. In such a case, the concepts cannot be completely

defined by a single operational definition. Researchers can review the literature and use operational definitions based on the definitions used in other
studies. In doing so, researchers have to make sure that the published definitions are applicable to their study, and that the reliability and validity of the
operational definition have been established. Another issue that needs to
be considered is that the operational definition should be independent of time
and setting (Burns and Grove 2005). Thus, the same definition of a variable
can be used at different times and in different settings. Lastly, the operational
definition is sometimes more restrictive and specific than the conceptual
definition. Care should be taken to make sure that broad concepts are not
"too narrowly interpreted."

Procedures for operationalizing concepts

Operationalization is a continuous process and involves a number of interrelated stages (Waltz et al. 2005). As each stage progresses, there is an ongoing
transformation from an abstract concept to a concrete one. The different
stages are not mutually exclusive and there is a lot of interplay between them.
The first stage involves developing the conceptual definition. The purpose is
to translate a vague concept and give it a precise meaning so that it can be
easily understood. Conceptual definitions can be developed using a number of
strategies including literature review, concept mapping, and developing theoretical models. An example of a multidimensional concept is "functional
status," which is defined as the ability of the person to perform activities of
daily living. The next stage is to identify the variable characteristics of the
concept. Variables are selected if they provide useful information about the
concept, help in understanding the concept, and are measurable. Several
dimensions of functional status can be identified including body care, exercises, social activities.

The next stage in the operationalization process is to select observable
indicators. For each dimension of functional status, specific observable indicators can be identified. For body care, indicators could be the ability to dress
including tying shoelaces and doing up buttons. The next stage is to develop
operational definitions of the indicators, i.e., developing means to observe
and measure indicators. The last stage is to evaluate if the concepts have been
appropriately and satisfactorily operationalized. This involves evaluating the
clarity, precision, reliability, validity, feasibility, and consistency of all variables and indicators used in defining a concept.

Summary and conclusions

In this chapter the four major steps in operationalizing a research study have
been discussed. Identification of a concise and clear research problem is
the first and most important step. Specific research questions and hypotheses

are then formulated to bridge the gaps between the research problem and study design, data collection, and analysis. Research questions focus on describing variables, examining interrelationships of variables, or evaluating causal relationships of single or multiple variables of interest. In contrast to research questions, hypotheses are specific statements of prediction and consist of at least one specific relationship between two variables, which are measurable.

Abstraction is a process of linking higher-level abstracts, which in turn are linked to the observable level through extensionalization. Operational definition, a method of extensionalization, is usually derived from a set of procedures or activities employed by the researcher in either manipulating an independent variable or measuring a dependent variable. The goal of operationalization is to move from a highly abstract concept to a concrete concept that is measurable.

In conclusion, formulating a clear and concise research problem is a challenging process. Research problems undergo several iterations until they are ready for empirical investigations. Based on the research problem, specific research questions and hypotheses are developed that give direction to the research study. As for research problems, research questions and hypotheses have to be refined and this requires a thorough search and understanding of the literature. Operational definitions are fundamental to an optimal research project and provide the bridge between incidental observation and scientific validation of theories and concepts. By spending thoughtful time on identifying and operationalizing a good research idea, a researcher can generate new knowledge that can go a long way to improving and augmenting current and future pharmaceutical practice and policy.

Review questions

1 Describe the approaches to conceptualizing a research problem.
2 Describe the characteristics of a good research question.
3 What are the different types of research questions and discuss their sources?
4 Define hypothesis and explain the different types of hypotheses in research studies.
5 What is operationalization and discuss the procedures for operationalizing research questions?

References

Behi R, Nolan M (1995). Deduction: moving from the general to the specific. *Br J Nursing* 4: 341–44.

Beitz JM (2006). Writing the researchable question. *J Wound Ostomy Continence Nursing* 33: 122–4.

Brink PJ, Wood MJ (2001). *Basic Steps in Planning Nursing Research. From question to proposal,* 5th edn. Sudbury, MA: Jones & Bartlett.

Burns N, Grove SK (2005). *The Practice of Nursing Research. Conduct, critique, and utilization,* 5th edn. Philadelphia: Elsevier Saunders.

Cummings SR (2007). Conceiving research questions. In: Hulley SB (ed.), *Designing Clinical Research.* Philadelphia, PA: Lippincott Williams & Wilkins.

DePoy E, Gitlin LN (1998). *Introduction to Research. Understanding and applying multiple strategies,* 2nd edn. St Louis, MO: Mosby, Inc.

Fain JA (2004). *Reading, Understanding, and Applying Nursing Research,* 3rd edn. Philadelphia, PA: FA Davis Co.

Gillis A, Jackson W (2002). *Research for Nurses: Methods and interpretation.* Philadelphia, PA: FA Davis.

Hepler C (1980). Problems and hypotheses. *Am J Hosp Pharmacy* 37: 257–63.

Kerlinger FN, Lee HB (2000). *Foundations of Behavioral Research,* 4th edn. Orlando, FL: Harcourt College Publishers.

Kwiatkowski T, Silverman R (1998). Research fundamentals: II. Choosing and defining a research question. *Acad Emerg Med* 5: 1114–17.

Lipowski EE (2008). Developing great research questions. *Am Soc Health-System Pharmacy* 65: 1667–70.

Meadows KA (2003). So you want to do research? 2: Developing the research. *Br J Commun Nursing* 8: 397–403.

Newton JT, Bower EJ, Williams AC (2004). Research in primary dental care Part 2: Developing a research question. *Br Dental J* 196: 605–8.

Norwood SL (2000). *Research Strategies for Advanced Practice Nurses.* Englewood Cliffs, NJ: Prentice-Hall, Inc.

Polit DF, Hungler BP (1991). *Nursing Research: Principles and methods,* 4th edn. Philadelphia, PA: Lippincott.

Stamler LL (2002). Developing and refining the research question: Step 1 in the research process. *Diabetes Educator* 28: 958–62.

Waltz CF, Strickland OL, Lenz ER (2005). *Measurement in Nursing and Health Research,* 3rd edn. New York: Springer Publishing Co.

Wood MJ, Ross-Kerr JC (2006). *Basic Steps in Planning Nursing Research: From question to proposal,* 6th edn. Sudbury, MA: Jones & Bartlett.

Online resources

Trochim WMK. (2006). "Structure of research." Available at: www.socialresearchmethods.net/kb/contents.php.

Institute of International Studies' Online Dissertation Proposal Workshop – Regents of the University of California. "The Research Question." Available at: http://globetrotter.berkeley.edu/DissPropWorkshop.

Johns Hopkins Bloomberg School of Public Health. Dissertation Workshop. Available at: http://ocw.jhsph.edu/courses/dissertationworkshop.

4

Measurement theory and practice

Rajender R Aparasu

Chapter objectives

- To explain the concept of measurement
- To discuss levels of measurement
- To describe reliability and methods to evaluate reliability
- To describe validity and methods to evaluate validity

Introduction

Measurement forms the basis for empirical health services research including pharmaceutical practice and policy research. The measurement process is designed to record and capture the underlying construct and concept. In healthcare, critical policy and practice discussions are based on cost, quality, and access. These concepts or constructs have to be measured and analyzed using variables to evaluate the performance of a healthcare system at the patient, family, institution, and population levels. Each of these constructs is complex and often requires multiple measures or variables to capture the underlying concepts. Advances in measurement of each of these dimensions have led to improvement in the healthcare system. For example, the development of the *Diagnostic and Statistical Manual of Mental Disorders* (DSM), diagnostic-related groups (DRGs), and Healthcare Common Procedure Coding System (HCPCS) has led to changes in healthcare delivery and reimbursement. In recent years, measurement of quality of healthcare based on structure, process, and outcome has gained significance due to its importance in healthcare delivery (Donabedian 2003).

The measurement process involves a systematic assignment of values or numbers to observations based on *a priori* rules of measurement (Viswanathan 2005). It is a critical step in quantitative research because it also defines the subsequent steps in conducting research such as analysis and interpretation of the research findings. The measurement process involves recording observations that are manifestations of the underlying construct; this requires a good understanding of the construct and the measurement process. Often the variables are operationalized based on the methodology used to capture these variables. This can involve collection of data by the researcher, also called primary data, or data collected by others for reuse, also referred to as secondary data. The operationalization and collection of data are critical in the measurement process. Further evaluation of this measurement process will confirm whether the process is truly measuring the construct, which entails ensuring that the measurement process is reliable and valid. This chapter describes the concept and levels of measurement, and discusses reliability and validity, and methods commonly used to evaluate reliability and validity.

Nature and level of measurement

The measurement process for any construct is based on the existing knowledge base regarding the construct. The measurement of physiological constructs such as blood pressure and blood glucose is often standardized. However, behavioral constructs such as compliance are rather complicated because the observed behavior may or may not reflect the intended construct. An understanding of the measurement process for behavioral constructs can provide a good framework to understand the measurement process in general. Summers (1970) suggests that the measurement process of an abstract construct such as behavior that is not directly observable involves three interlinked operational subprocesses:

1 Identification of acceptable behavior specimens that represent the underlying construct
2 Data collection of specimens
3 Conversion of specimens to a quantitative variable.

An acceptable specimen defines the data collection process, which in turn determines the type of quantitative variable. Often operational definitions, especially measured operational definitions, are based on these three operational subprocesses.

Acceptable behavioral specimen

Identification of acceptable specimens that represent the underlying construct is a critical step in the measurement process. The current knowledge base defines ones that are acceptable and those that are not. Acceptable specimens for diagnosis of diabetes such as blood glucose and glycated hemoglobin are often standardized. The measurement process for constructs such as compliance is complex. However, the concept of compliance with a medication regimen is important in evaluating treatment effects. Compliance refers to the extent to which the patient follows healthcare advice (Hayes et al. 1979). Not all patient behaviors that reflect compliance with a medication regimen are easily observable by researchers and clinicians. Compliance with a medication regimen can include behaviors that can capture the extent to which the patient is taking medication with respect to frequency and duration of therapy. The behavior specimens such as refill history which capture frequency and duration of therapy are considered acceptable for measuring compliance (Farmer 1999). Other medication-taking behaviors can include time of administration and avoidance of certain foods.

Data collection

The data collection process entails capturing data based on specimens. Often multiple data collection processes can be linked to the behavior specimens. With respect to compliance, the data collection can involve self-reports of specimen behavior or observation of overt behavior such as refill history (Farmer 1999). The self-report methodology involves measurement of patient responses to a series of questions or items using a survey instrument to determine medication-taking behaviors. Self-report methodology is based on the assumption that the patients can report their medication use behavior. Overt behaviors of compliance can be assessed by direct observation of patient behavior or refill history as captured in prescription claims data. Refill history measures, such as medication possession ratio, reflect the number of doses filled by the patient for the dispensing period. Refill history is based on the assumption that prescription-filling behavior reflects the patient's medication use pattern. Compliance measurement based on biochemical instruments to evaluate drug levels can also be used as a behavior specimen (Farmer 1999).

Assignment of values

Converting the behavioral specimen to a variable involves assignment of values based on the rules developed for data collection to evaluate the underlying concept. The assignment process should reflect variation in the underlying construct. For example, the value assigned to a compliance measure, based on self-reports or refill history, should reflect the extent to which the patient is following the healthcare advice for medications. Measurement and further analysis are dependent on the properties of the variable. The levels of measurement can be classified as nominal, ordinal, interval, and ratio, based on the properties of the quantitative variable. This is also referred to as Stevens's (1946) scales of measurement.

The nominal measure, also called the categorical measure, is the simplest and lowest form of measurement. It is used for naming or identification purposes. Examples include measurement of gender (male or female) and ethnicity (white, black, etc.). The subgroups in a nominal variable are mutually exclusive and exhaustive. All members of a group have the same identity and there is no relationship among the subgroups. For nominal measures the assignment of numbers, letters, or symbols is only for labeling and grouping purposes. Although numbers are often used for labeling or coding purposes, they do not have an arithmetic value. Consequently, none of the mathematical manipulations can be used for nominal measures. The nominal measures are often used for counting and to examine frequency distribution. The number of subgroups for classification or labeling is based on the extent of identification needed. For example, the respondents' residences can be grouped by state (50 states) or by region (northeast, midwest, south, and west). Nominal variables can be used as dichotomous variables where only two subgroups are recorded such as white and nonwhite for race.

The ordinal measure is rank ordering of a group membership with properties of transitivity. The group membership can be ordered from lowest to highest with a clear interrelationship of the levels, unlike the nominal measure. For example, the health status of a patient can be measured with levels or responses of excellent, very good, good, fair, or poor. Similarly, a patient's perception of compliance can be measured using a five-point scale from most compliant to least compliant. Ordinal measures are often used in survey instruments, and can be ordered from highest (excellent) to lowest (poor). In addition, the rank order relationship levels are mutually exclusive and exhaustive, similar to the nominal measure. In addition the interrelationship of the levels is known. The distance between the levels is, however, not equal and, at times, not known. The levels have properties of transitivity, i.e., if excellent is better than very good and very good is better than good, this makes excellent better than good.

The interval and ratio measurements have properties of equality of intervals in addition to characteristics of an ordinal measure. The key difference between interval and ratio measures is that there is an absolute zero level in ratio measure. Ratio is the highest form of measurement because it can represent the zero value of the measure. Mathematical manipulations such as multiplication and division can be used for ratio measures. Examples of ratio measures include prescription expenditure, hospitalization days, wait time, and weight gain. The best example of an interval measure is temperature measured on the Fahrenheit scale. The temperature is measured using the equidistant scale of Fahrenheit in which there is no zero amount or quantity of temperature; the assignment of zero temperature is arbitrary. The intelligence quotient (IQ) scale is another example of an interval measure where the distance between the scale values is the same. In psychometric research, it is difficult to find a measure that has an absolute zero. The interval measures allow mathematical manipulations. In general, ratio and interval measures are used in the same way for statistical analysis. Ratio and interval measures provide great flexibility for mathematical manipulations as long as the original properties of measurement are satisfied. With respect to compliance, the number of missed doses is a ratio scale that can be obtained from patients using self-reports.

Measurement issues

Ratio and interval measurements are the most desirable measures in health services research because they capture more information about the underlying construct, and are therefore more likely to reflect the variation in the underlying construct, than other ordinal or nominal measures. Ratio and interval scales also have all the necessary properties required for mathematical manipulations. Most importantly, these measures can be converted to low forms of measures such as ordinal measures. It is not possible to convert ordinal measures into ratio/interval measures. For example, family income captured in dollars (ratio measure) can be grouped into high, middle, or low income (ordinal measure) based on specific ranges of family income. However, it is not possible to convert family income captured as an ordinal measure into a ratio measure. In general, higher forms of measures should be preferred to lower forms of measures.

The goal of the measurement is to assign values to a variable based on the specific rules formulated to measure the underlying construct or concept. The measurement process is also designed to capture variation in the underlying construct. The values assigned in the measurement process are distinct due to the differences in the underlying construct or concept. If variations in the underlying construct or concept are not reflected in the measurement, it leads to measurement error, which is an error in the measurement process (Viswanathan 2005). Sources of measurement error can be due to the measurement process itself or factors outside the measurement process. For example, a measurement error may occur if the number of missed doses is captured using self-reports in the presence or absence of a physician. A good measurement process is designed to reduce such measurement errors. These issues are considered in detail in the discussion of reliability and validity.

Measurement and statistical analyses are based on the concept of analysis of the variation. The type of measurement of variables determines the statistical analysis to be used. This holds true for both dependent and independent variables. The interval and ratio measures capture greater variation in the dependent measures than the ordinal measures. Consequently, there is more opportunity to explain the extent of variation in interval and ratio measures than ordinal measures. This can also improve to capture the sources of variation in statistical analysis.

Reliability and validity

The goal of the measurement process is to ensure that the values assigned to variables are reliable and valid. Reliability and validity are different dimensions of the measurement process. Reliability ensures that the assignment of values is consistent or reproducible, whereas validity ensures that the assignment of values truly reflects the underlying construct or concept (Bohrnstedt 1970; DeVellis 1991; Trochim 2001). Both reliability and validity are important in the measurement process because reproducibility of a measure as well as the trueness of a measure is critical in research. Reliability of a measure does not, however, ensure its validity. For example, use of self-reports to capture compliance with medication may be reliable but may not be valid. Self-reports of missing doses may consistently provide the same measures but may not truly capture the patient behavior. The constructs in sociobehavioral research are often abstract and hence require evaluations of reliability and validity of the measurement process. Psychometric research has played a significant role in the evaluation of reliability and validity. Some of the reliability and validity evaluation methods are specific to survey research involving a survey instrument. However, the concepts of reliability and validity are relevant for both behavioral and nonbehavioral constructs.

Measurement errors

Both reliability and validity of a measurement are affected by measurement errors. Measurement errors can be classified as random errors and nonrandom errors (Viswanathan 2005). Random or chance errors, as the term suggests, occur inconsistently, and cause the measures to deviate randomly from the true value. Random errors negatively affect the reliability of the measurement and are present in every measurement process. The goal of a measurement process is to minimize random errors and, thereby, maximize the reliability. The factors that influence reliability can be the individual, instrumentation, and environment. Individual or patient level factors include diurnal variation, education level, or biological variability. Instrument level factors include calibration of instrument, or misreading or mistakes in recording questionnaire responses. Environmental factors influence individual and instrumental factors, such as temperature, pressure, light, or electrical fluctuations. An ideal measurement process will minimize the influence of these factors to maximize the reliability. This can be achieved using standard and consistent data collection and administration methods. For example, the measurement can be recorded at one specific time for all participants to minimize diurnal variation.

Nonrandom or systematic errors occur consistently by definition and hence cause the measures to deviate from the true value nonrandomly (Viswanathan 2005). The amount of systematic error directly influences the validity of the measurement. It is inversely related to validity: high nonrandom error decreases the validity of a measurement process. Systematic bias is a classic example of a nonrandom error that threatens the validity of a measurement process. Random bias does not influence the validity of a measurement.

There are different types of biases such as information, recall, and interviewer bias. Information bias occurs when there is consistently differential information among the participants of interest due to underlying factors. For example, in epidemiological research, test cases tend to provide more information than control cases that do not have the disease. Recall bias occurs where there is a differential ability to recall information about previous experience due to issues related to time or experience. For example, the measurement error in reporting the number of missed doses is likely to be less than the number of doses actually taken according to the instructions due to recall bias. Interviewer bias exists when interviewer perception or behavior influences the responses. The measurement process should preferably control for systematic biases at both data collection and study design stages to strengthen the validity. For example, blinding techniques are used to hide the procedural aspects of study design from respondents and interviewers to minimize the bias.

Reliability

Reliability, as discussed earlier, ensures that the measurement is consistent or reproducible (Bohrnstedt 1970; DeVellis 1991; Trochim 2001). A reliable measurement process will consistently provide the same or a similar value for an unchanged variable, whereas a change in the underlying construct will reflect a change in the value assigned. According to Bohrnstedt (1970), the reliability assessments can be grouped into two major classes: measurement of stability and measurement of equivalence. Measures of stability evaluate relationship or correlation of measures across time or evaluators, and examples include the test–re-test method and interrater reliability. Measures of equivalence evaluate relationship or correlation between two sets of instruments, and examples include split-half, parallel form, and internal consistency methods.

Measures of stability

The test–re-test method analyzes measures obtained from the same participants across time using the same instrument. It evaluates the stability of measurements over time. The instrument for measurement can be any equipment such as a weighing scale or a survey instrument with a series of questions. For example, weight measured using a weighing scale at one time can be correlated with weight obtained after 1 hour. The time difference between the measurements is dependent on the type of measure. A longer time interval can influence the reliability for some measures due to changes in the underlying construct such as a person's weight. For other measures involving examinations, shorter time can influence the reliability due to knowledge of the previous test. The general rule is that the time interval should be long enough that respondents do not remember their responses without the change in the underlying construct. The correlation between the two measures on the same participants provides the correlation coefficient. This method is often used to ascertain the reliability of physiological measurements such as blood glucose and blood pressure. In fact, diagnostic criteria for a disease are often based on the test–re-test method. Pearson's correlation is used to calculate the correlation coefficient for interval–ratio measures, whereas Spearman's rank correlation is used for ordinal measures. In general, the correlation coefficient of >0.80 is considered as a reliable determinant (Nunnally and Bernstein 1994). Disadvantages of the test–re-test method include inconvenience and reactivity. Due to multiple measurement processes it is inconvenient to participants and researchers. Reactivity refers to change in the underlying construct due to testing. For example, patients responding to a question related to compliance are sensitized to the issue of compliance and thus provide responses that might reflect an improvement in compliance.

Interrater reliability involves analysis of measures obtained from the same participants by different evaluators using the same instrument. For example, pharmacists' counseling time can be measured by two independent observers. The reliability of such measures can be evaluated by correlating the time measures obtained from the two observers. Cohen's κ coefficient is used for reliability involving two evaluators, whereas Fleiss's κ is used for measures involving multiple evaluators (Landis and Koch 1977). These measures are calculated based on the difference between percentage of agreement among evaluators and probability of chance agreement.

Measures of equivalence

The split-half method is one of the earliest measures of equivalence to determine questionnaire reliability. This method involves dividing the number of items or individual questions of a survey instrument into two equivalent halves; the correlation between the two halves provides the correlation coefficient. There are two options for dividing the items or questions: one method involves dividing the items into even and odd questions; the other involves dividing the items into first and second halves. The decisions on the type of splitting are based on practical considerations and type of items in the survey instrument. Irrespective of the approach, the split-half method provides two measures on each participant based on two equivalent-form measures, and the measures from the halves are then used to calculate the correlation coefficient. The basic underlying principle in the split-half method is that the two halves are designed to measure the same underlying construct. The strength of the split-half method is that it overcomes the problems of test–re-test methods such as reactivity and inconvenience. The weakness of the split-half method is that two halves must measure the same underlying construct and the reliability coefficient can vary based on the approach used to divide the items. The Spearman–Brown prophecy formula is generally used to obtain the correlation coefficient between the split halves (DeVellis 1991). The formula is based on the correlation coefficient between the split halves and correction needed to divide the items in the survey instrument into two halves.

The parallel-form method is an extension of the split-half method in which two parallel questionnaires are administered to each participant consecutively. The time interval between administration of the two questionnaires should be minimal to optimize changes in the underlying constructs. The scores from the two forms or questionnaires are used to calculate the correlation coefficient. Similar to the split-half method, the two forms should be equivalent and measure the same underlying construct but should not be identical. This method also overcomes the problems of test–re-test methods such as reactivity and inconvenience. However, it may be cumbersome to the respondents to complete two parallel questionnaires. In general it is easier to develop similar items in the split-half method than to create parallel forms. Consequently, the parallel-form method is seldom used in health services research but often used in educational research involving examinations due to experience in creating parallel exams and availability of a large pool of questions.

The internal consistency method is the most frequently used method in health services research. It involves correlation among all items or questions in a questionnaire without the need to divide items or create forms. The internal consistency method evaluates whether all items in a questionnaire are measuring the same construct. It also overcomes problems associated with split-half and parallel-form methods, such as varying reliability due to the process used to divide the items or create the forms. Internal consistency is based on the concept that items or questions designed to measure the same underlying construct should be highly correlated. This means that each item or question is used to compare consistency of responses with other items in the questionnaire for the study sample. As a result, the correlation coefficient is sample specific. There is a need to assess the reliability of the survey instrument with a change in the study sample. There are several ways to compute a reliability coefficient based on the internal consistency approach. Cronbach's α is used to calculate the internal consistency of measures based on continuous measures (Bohrnstedt 1970; DeVellis 1991). The reliability coefficient increases with an increase in the number of items and inter-item correlations. The Kuder–Richardson coefficient (KR20 or KR21) is used to calculate reliability for nominal measures. These coefficients are calculated based on the proportion of same responses for an item. High conformity of responses leads to high KR20 or a reliability coefficient of nominal measures. Cronbach's α is an extension of KR20 and both calculations are based on classic test theory. Most statistical packages can calculate these reliability coefficients.

Validity

Validity, as mentioned earlier, ensures that the instrument developed for measurement purposes truly represents the underlying construct (DeVellis 1991; Nunnally and Bernstein 1994; Trochim 2001). In addition to appropriateness, instruments are increasingly being evaluated for meaningfulness and usefulness, in recent years, in order to strengthen the validity. Although methods to ascertain validity have changed over the years, construct validity has remained the cornerstone of all types of validity assessments. Construct validity refers to the degree to which the instrument measures the underlying construct. The evidence to strengthen the construct validity is based on internal structure and external relationships. The internal structure evaluates the interrelationship of the items and underlying dimension of the construct. The external relationship evaluates the relationship of the instrument to other constructs.

The internal structure of the instrument should be consistent with the underlying dimensions of the construct. Factor analysis is commonly used in the development process of a survey instrument (DeVellis 1991; Nunnally and Bernstein 1994). Exploratory factor analysis helps to identify various factors or dimensions that are represented in the instrument. It also groups items that belong together representing the underlying construct. For example, this can be used to establish the dimensions of a quality-of-life scale. Exploratory factor analysis is usually followed by confirmatory factor analysis to determine the extent to which statistical validity is based on the underlying theoretical model. Although factor analysis approaches are useful, the internal structure should not be the only basis of construct validity. The relationship to other constructs also ascertains validity of the construct. External relationships should be empirically tested for hypotheses developed based on the theoretical relationships. This addresses the evidence of predictability and discernibility.

Trochim (2001) proposed that all types of validity testing methods should strengthen the construct validity of an instrument. This can be achieved using translational and criterion validity. The translational validity addresses the translational or transformational aspect of construct validity, which includes face and content validity. These validity analyses are designed to ensure that the items in the instrument reflect the underlying construct. Face and content validity will ensure that the items represent the intended factors or dimensions of the construct, which can also be confirmed using exploratory factor analysis. Criterion validity refers to the relationship aspect of construct validity, and includes concurrent, predictive, convergent, and discriminant validity. Methods for criterion validity empirically test for theoretical relationships. The translational validity and the criterion validity ensure appropriateness, meaningfulness, and usefulness of the instrument. Multiple methods are needed to strengthen the construct validity of the instrument.

Translational validity

Face validity is the simplest method to ensure translational validity. It addresses the question: Do these items and the overall instrument measure the underlying construct? This involves a judgment or an opinion of a layperson or an expert. A layperson will provide his or her perspective, mainly to address the issues related to the interpretation of items and administration of the instrument. Experts can provide detailed opinion about the appropriateness and wording of items, organization of items, and the overall look of the instrument with respect to the underlying construct. It primarily involves qualitative and subjective assessment of the instrument. Consequently, it is considered as the weakest form of validity assessment. It is often used in the development process to refine an instrument.

Content validity refers to the representative nature of the items to capture the underlying dimension of the construct. It presents the relationship of items to the dimensions of the underlying construct. The content validation process requires a clear definition of the dimensions of the underlying construct and ways to ensure that the selected items represent the relevant dimensions. The dimensions of the construct can be defined from the literature or expert opinion. For example, measurement of healthcare quality requires items or questions related to the structure of the healthcare system, process of obtaining healthcare, and outcomes of healthcare obtained (Donabedian 2003). A content expert's opinion can be sought to evaluate whether the measurement items represent the defined dimensions. Although there is some subjectivity in the process, seeking the opinions of multiple experts can reduce subjectivity and improve the face validity. Also, analytical measures such as content validation ratio, content validation form, and content validation index can strengthen the content validity (Shultz and Whitney 2005). Content validity will ensure that the items and the overall instrument reflect the dimensions of the construct.

Criterion validity

Criterion validity addresses the relationship aspect of the construct validity by attesting to the relationships between the instrument and criterion, or other measures, based on theory and practice. Selection of the criterion plays an important role in criterion validity. The evidence and the extent of the inter-relationship of criteria strengthen or weaken the construct validity. A strong relationship means that the criterion is well validated and accepted. Often the criterion selected is external and considered the "gold standard." The theoretical and practical knowledge about the issues are critical in selection of the criterion. For example, compliance measured using self-reports or refill history can be validated using the criterion of drug levels in blood or urine because these are considered the gold standard. As mentioned earlier, types of criterion validity include concurrent, predictive, convergent, and discriminant validity.

Concurrent validity is a type of criterion validity that refers to the relationship between the instrument and the criterion measured at the same point in time. The criterion selected for concurrent validity should measure the same underlying construct as that of the instrument. The selected criterion should be a standard measure. The rationale for concurrent validity is that, if the instrument and the criterion are administered at the same time and measure the same underlying construct, then there should be strong correlation between the two measures. For example, compliance measured using self-reports can be validated by comparing responses with the drug levels in blood or urine measured at the same time.

Predictive validity is a type of criterion validity that addresses the relationship between the instrument and criterion measured at a future time. The criterion and the instrument are not measuring the same underlying construct as in the concurrent validity. However, the instrument should be able to predict the criterion. For example, compliance measures based on refill history have been shown to predict the health expenditure in patients with diabetes. This is based on the hypothesis that disease is managed better in compliant than in noncompliant patients, and thus leads to a decrease in healthcare expenditure. The rationale for predictive validity is that, if the patients with diabetes are compliant, they will incur less expenditure due to better disease-state management. Therefore, the criterion selected for predictive validity should be based on theory and practice.

Convergent and discriminant validity are two sides of the same concept. Convergent validity refers to convergence or a strong relationship between the instrument and the criterion, which are theoretically similar. Discriminant validity refers to little or no relationship between the instrument and the criterion, which are theoretically different. Convergence validity is similar to concurrent validity but it is not restrictive with respect to time of administration. The concept of convergent and discriminant validity is based on the principle that, if the instrument is valid, it will be strongly correlated with measures that are similar and will not be associated with measures that are dissimilar. For example, instruments to measure pain and overall quality of life will be strongly associated because they measure similar concepts. Conversely, pain measures are less likely to be associated with perceptions on economy as these are dissimilar concepts.

Measurement process and practice

Understanding the measurement theory and practice is vital in conducting empirical research. An existing knowledge base defines the methodology to operationalize a construct, which includes identification of acceptable behavior specimens, data collection, and assignment of values to develop the variable for an underlying construct. Some constructs are easy to measure, such as expenditure; other constructs, such as quality, need a strong understanding of underlying theory in order to operationalize. Prescription expenditures are usually captured using secondary data sources such as claims data. The behavior specimen is reflected in claims data and it captures payment by insurance companies and other sources. The data collection involves use of secondary data and value assignment of expenditures involves a ratio scale. Other constructs such as quality are complex and require significant effort. The existing knowledge base suggests that quality measurements should be based on the underlying dimensions. For example, quality of medical care is based on measures of structure, process, and outcome (Donabedian 2003). The operational definitions and measurements processes for each of these dimensions are different. According to Donabedian (2003), measures of quality are relevant only when there is an interrelationship for structure, process, and outcome. Consequently, greater understanding of the underlying theory and advances in data collection methodologies play an important role in the measurement process and operationalization of healthcare constructs.

Reliability and validity issues are critical in the identification of behavior specimen and data collection phases. Concepts of reliability and validity are valuable to develop and improve a strong research instrument. Accordingly, they should be considered tools for continuous quality improvement of the measurement process. A well-defined construct helps to ensure the construct validity of an instrument, which includes translational and criterion validity. Translational validity includes face and content validity. The considerations of face and content validity can be incorporated in the identification of the behavior specimen phase to ensure that instrument development is consistent with the underlying dimensions of the construct. The criterion

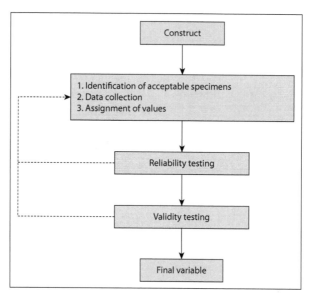

Figure 4.1   Measurement process and practice.

validity is difficult to incorporate into the behavior specification phase. In general the considerations of translational validity are likely to ensure criterion validity. Criterion validity can be tested only by correlating measures from the instrument with a criterion. The test findings will reveal needed improvements in the development phase. Figure 4.1 provides a schematic diagram for the measurement process and practice.

In the data collection phase, the developed instrument is utilized to collect the data. Various research techniques can be employed to ensure reliability and validity of data collection. Standardized data collection and administration methods can minimize random errors and improve reliability. Pre-testing of the instrument can help to identify items that require clarification. It can also help to improve the organization of the instrument. Response biases can be minimized using pre-tested items, techniques of blinding, utilization of trained interviewers, and consistent data collection methods. The data collection process should minimize random errors and control for nonrandom errors to maximize reliability and validity. The tests for reliability and validity will ensure that the instrument developed and utilized for research is reliable and valid.

Summary and conclusions

The measurement process is designed to record and capture the underlying construct. This involves identification of acceptable specimens, data collection of specimens, and conversion of specimens to a quantitative variable. The decisions made at each of these interrelated steps are based on the existing knowledge base of the construct. Reliability and validity issues are critical in the measurement process. Reliability addresses stability and equivalence of the measurement process. The tests of translational and criterion validity are designed to ensure construct validity of the measurement process. Construct validity addresses the extent to which the variable measures the underlying construct. A reliable and valid measurement process will minimize measurement errors and thereby strengthen the research. Measurement also forms the basis of subsequent steps in research such as statistical analysis.

Review topics

1 Discuss levels of measurement using examples.
2 Describe the concept of reliability and methods to evaluate reliability.
3 Discuss common types of measurement errors.
4 Describe the concept of validity and methods to evaluate validity.
5 Describe the measurement process using an example in pharmaceutical practice and policy research.

References

Bohrnstedt GW (1970). Reliability and validity assessment in attitude measurement. In: Summers GF (ed.), *Attitude Measurement*. Chicago, IL: Rand McNally, 80–99.

DeVellis RF (1991). *Scale Development: Theory and applications*. Thousand Oaks, CA: Sage Publications.

Donabedian A (2003). *An Introduction to Quality Assurance in Health Care*. New York: Oxford University Press.

Farmer KC (1999). Methods for measuring and monitoring medication regimen adherence in clinical trials and clinical practice. *Clin Ther* 21: 1074–90.

Hayes RB, Taylor DW, Sackett DL (1979). *Compliance in Health Care*. Baltimore, MD: Johns Hopkins University Press.

Landis JR, Koch GG (1977). The measurement of observer agreement for categorical data. *Biometrics* 33: 159–74.

Nunnally JC, Bernstein IH (1994). *Psychometric Theory*, 3rd edn. New York: McGraw-Hill.

Shultz KS, Whitney DJ (2005). *Measurement Theory in Action: Case studies and exercises*. Thousand Oaks, CA: Sage Publications.

Stevens SS (1946). On the theory of scales of measurement. *Science* 103: 677–80.

Summers GF (1970). Introduction. In: Summers GF (ed.), *Attitude Measurement*. Chicago, IL: Rand McNally, 1–21.

Trochim WMK (2001). *The Research Methods Knowledge Base*, 2nd edn. Cincinnati, IL: Atomic Dog Publishing.

Viswanathan M (2005). *Measurement Error and Research Design*. Thousand Oaks, CA: Sage Publications.

Online resources

Agency for Healthcare Research and Quality (AHRQ). "National Quality Measures Clearinghouse." Available at: www.qualitymeasures.ahrq.gov.

Centers for Disease Control and Prevention (CDC). National Center for Health Statistics. Series 2. "Data Evaluation and Methods Research." Available at: www.cdc.gov/nchs/products/series.htm#sr2.

Centers for Disease Control and Prevention (CDC). National Center for Health Statistics. Series 6. "Cognition and Survey Measurement." Available at: www.cdc.gov/nchs/products/series.htm#sr6.

The Leapfrog Group. Available at: www.leapfroggroup.org.

5

Experimental designs

Kenneth A Lawson

Chapter objectives

- To explain criteria for establishing causal relationships
- To describe threats to internal and external validity
- To describe common experimental designs
- To identify strengths and weaknesses of experimental designs

Introduction

According to Kerlinger and Lee (2000), research design describes the plan and structure of a study to answer a research question. An important step in the research planning process for any study is determining which research design is appropriate. That decision depends on several considerations including the nature of the research questions to be addressed. For studies that seek to determine whether causal relationships exist between variables, experimental research designs are preferred because their characteristics can control for factors that may confound those relationships.

This chapter begins with a discussion of the nature of causal relationships and the criteria for establishing causality. Next, the characteristics and types of experimental research designs are introduced, followed by a description of the threats to internal and external validity which are critical considerations in selecting the appropriate design. Selected experimental designs are presented and discussed with respect to their control of validity threats. The final section of the chapter covers the strengths and limitations of experimental designs. Although the concepts discussed in this chapter are applicable to research in any topic area, they are presented as they relate to pharmaceutical practice and policy research.

Causal relationships

Researchers sometimes seek answers to questions that are descriptive in nature (e.g., How many new cases of influenza were reported last year?). But they often address questions about cause and effect. Although it may seem reasonable that some presumed cause is responsible for an observed effect, other factors (both obvious and subtle) may be influencing outcomes. Experimental research can provide evidence to help determine if hypothesized causal relationships truly exist.

Polgar and Thomas (1991) propose three important criteria for the demonstration of causality:

1 Antecedent occurrence of cause to effect
2 Covariation of cause and effect
3 Elimination of rival cause explanations of the effect.

Antecedent occurrence means that a cause must occur before an effect occurs. A study testing the hypothesis that preservatives in vaccines cause autism must show that autism did not exist before vaccine exposure. Covariation means that an empirical relationship must exist between the hypothesized cause and effect. Covariation exists if people who receive vaccines with preservatives are more likely to develop autism compared with those who receive preservative-free vaccines. To meet the third criterion, elimination of rival cause explanations, alternate explanations of the effect must be eliminated. In this example, the study must account for other potential causal factors that may be present in people who receive vaccines with preservatives.

Polit and Beck (2008) propose another important criterion specific to health research: the presence of biologic plausibility. The causal pathway must be credible based on laboratory or basic physiologic studies. Causal relationships in the physical sciences are (arguably) easier to establish because they obey physical laws. For example, combining the same reagents in the same amounts under the same conditions will always yield the same results. However, causal relationships in the health, social, and behavioral sciences are more difficult to establish because of variations among people (participants) and lack of control over study conditions. Various study designs and analytic approaches have been developed to address these issues.

Experimental research

Research designs may be nonexperimental, quasi-experimental, or experimental. All three types can provide valuable information when used appropriately. However, experimental research designs are preferred over other designs for studying causal relationships because their characteristics are better suited to meeting the criteria for demonstrating causality (Shi 2008).

Design characteristics of controlled experiments

The "gold standard" for studying causal relationships is the controlled experiment (also known as the randomized controlled trial or randomized clinical trial [RCT] in medical research). The three major characteristics of controlled experiments are (1) manipulation, (2) control, and (3) randomization (Polit and Beck 2008). These design characteristics are responsible for the strength of controlled experiments in studying causal relationships.

Experimental studies examine causal effects of a treatment or other type of intervention (the primary independent variable) on designated outcomes (dependent variables). Manipulation refers to the researcher administering an intervention to some study participants. Interventions may take various forms: in clinical drug trials, the intervention is the active drug; in medical studies, the intervention is some form of treatment (e.g., a surgical procedure or a radiation treatment); and, in health services research, the intervention could be a wellness program. All independent and dependent variables (including the intervention) in an experiment should be operationally defined in a specific and explicit manner (Shi 2008). Also, the intervention must be administered to participants consistently throughout the study to ensure study validity.

Control refers to the control condition introduced by the researcher in the form of a control group of participants not receiving the intervention. Campbell and Stanley (1963) noted that at least one comparison is needed to obtain scientific evidence. In experimental studies, that comparison is between the experimental and the control groups. The experimental group receives the intervention (e.g., active drug, surgical procedure, or educational program) whereas the control group does not (or, in some studies, receives usual care, a different intervention, or a placebo). The outcome variables are measured for the experimental and control groups before (pre-test) and after (post-test) the intervention is administered. Comparison of the outcome variables allows researchers to examine the experimental group for the effects of the intervention and the control group for any changes in outcomes that occur in the absence of the intervention. Assuming that other research design characteristics control for confounding factors, differences between groups seen from this comparison can be attributed to the intervention.

Randomization or random assignment means that the researcher randomly assigns participants to the intervention group or the control group. As every participant has an equal chance of being placed in any group, random assignment should result in groups that are equivalent (or at least differ only by chance) with respect to participants' characteristics (known and unknown) that could influence study outcomes. This allows researchers to be confident that any observed differences between groups will be due to the intervention rather than to other factors; it is the most effective way of eliminating alternate explanations of the outcomes (Shi 2008). Random assignment is not always feasible or practical (Saks and Alsop 2007), and it does not guarantee group equivalence; therefore, equivalence should be confirmed by comparing groups on relevant baseline characteristics. In studies that do not use random assignment (e.g., quasi-experimental studies), the control group is termed a "comparison group." Researchers generally attempt to establish comparison groups that are as similar to the experimental groups as possible in these studies. Some studies use multiple experimental and control groups.

Other characteristics of controlled experiments

In addition to the three major characteristics noted above, some controlled experiments (particularly clinical trials) use blinding and inclusion/exclusion criteria as control mechanisms.

Blinding

Participants may bias their responses and researchers/observers may bias their observations (intentionally or unintentionally for both) (Cummings et al. 2001). For example, participants in a clinical drug study who know that they are receiving the active drug may be more responsive than those who know that they are receiving a placebo. Similarly a researcher who developed a medication may be more likely to see improvement in the condition of the participants who received the medication than in those who received the placebo. To minimize these potential biases, studies are sometimes blinded. In single-blinded studies, participants are unaware of their group membership, but researchers/observers are aware. In double-blinded studies, neither participants nor researchers/observers are aware of the participants' group membership. Although blinding is widely used in clinical drug trials to reduce bias, it is not feasible in many studies (e.g., a study of a pharmacist-run diabetes education program where participants and researchers are both likely to be aware of the participants' group membership).

Inclusion and exclusion criteria

Inclusion and exclusion criteria are used to specify the characteristics of the target population of the study. Participants are selected into the study sample so that their characteristics conform to the inclusion and exclusion criteria. To the extent that the study participants' characteristics match those of the target population, generalizability of the study findings is enhanced. Grady et al. (2001) identify factors that should be considered in establishing inclusion and exclusion criteria for clinical studies. Inclusion criteria are generally based on demographic, clinical, geographic, administrative, and temporal (study timeframe) characteristics that are relevant to the study. Exclusion criteria are used to identify individuals who will not be studied because of the following: their characteristics do not match those of the target population; a high likelihood of being lost to follow-up; an inability to provide good data; being at high risk for side effects; and characteristics that make it unethical to withhold the study treatment. Other factors may be relevant depending on the nature of the study.

Types of experiments

Several types of experiments are available for researchers including laboratory experiments, field experiments, natural experiments, and simulation experiments.

Laboratory experiments

The terms "controlled experiment" and "laboratory experiment" are often considered to refer to the same type of study. Shi (2008) defines laboratory experiments as "experiments conducted in artificial settings where researchers have complete control over the random allocation of subjects to treatment and control groups and the degree of well-defined intervention." They represent the ideal, "gold standard" type of experiment. They are widely used in medical research (including clinical drug trials) where they are known as RCTs. However, they are rarely used in health services research because of difficulties in implementing the experimental design and achieving other experimental conditions.

Field experiments

Field experiments are studies conducted in realistic settings where the experimenter manipulates the intervention and has as much control over the conditions as is feasible (Kerlinger and Lee 2000). Many health services, educational, and social science research studies occur outside the laboratory in realistic settings. Although researchers maintain some control over the interventions in field experiments, they typically have less control over the interventions and other experimental conditions compared with laboratory experiments. The Rand Health Insurance Experiment is a well-known example of a field experiment in health services research (Rand Health 2008). The goal of this study was to determine how variations in patient cost-sharing amounts affected health services utilization and patient outcomes. Patient cost-sharing amounts (the intervention) were manipulated by the researchers who studied how different cost-sharing amounts affected participants' utilization of healthcare services in a realistic setting.

Natural experiments

Shadish et al. (2002) describe a natural experiment as a "naturally occurring contrast between a treatment and a comparison condition." Researchers do not control interventions in natural experiments; therefore, these are nonexperimental studies. However, they provide opportunities to study the effects of interventions that occur naturally or that are controlled by people other than the researchers. An example of a natural experiment is a study conducted by Soumerai et al. (1991), which compared admissions to hospitals and nursing homes among Medicaid patients in two north-eastern states, one of which implemented Medicaid drug-payment limits and the other of which did not. Although not a true experiment because the researchers did not manipulate the intervention (drug-payment limits), this study contributed important insight to the effects of limits on drug payments in a natural setting. Natural experiments generally have high external validity and limited internal validity.

Simulation experiments

Shi (2008) describes simulation or modeling research as "a special type of experiment that does not rely on subjects or true intervention." Simulation research uses specific analytic approaches such as decision analysis modeling or Markov modeling to project the outcomes that will be experienced under specified conditions over time given specific inputs. For example, St Charles et al. (2009) conducted a Markov modeling study to evaluate the projected long-term (60-year) cost-effectiveness of continuous subcutaneous insulin infusion compared with insulin via multiple daily injections in patients with type 1 diabetes. The researchers estimated treatment costs, changes in glycated hemoglobin (HbA1c), progression and costs of disease-related complications, and discount rates in developing their model. Compared with controlled experiments, simulation studies are less costly, take less time to complete, do not expose patients to potentially harmful conditions, and allow researcher control over the model structure and input variables; however, a disadvantage is their artificiality (Shi 2008).

Validity considerations in experimental research

Shadish et al. (2002) refer to validity in general terms as the "truth of an inference." For a given study, the research design, the data source and quality, the data analyses and results, and the conclusions drawn from the results must be considered by those making a judgment about whether or not the hypothesized causal relationship truly exists. They proposed a typology consisting of four components: internal validity, external validity, statistical conclusion validity, and construct validity. Factors related to these components are considered in determining overall study validity. Threats to internal and external validity are discussed here; statistical conclusion validity and construct validity are addressed in other chapters.

Threats to internal validity

Polit and Beck (2008) define internal validity as "the extent to which it is possible to make an inference that the independent variable is truly causing or influencing the dependent variable and that the relationship between the two is not the spurious effect of a confounding variable." As the goal of experimental research is usually to determine the cause-and-effect relationship between an independent variable and a dependent variable, internal validity is an important consideration. In experimental studies, researchers must use research designs and methods that control the threats to internal validity to the greatest extent possible. Commonly recognized internal validity threats include selection, history, maturation, mortality/attrition, testing, instrumentation, and statistical regression.

Selection

Selection bias may occur when participants self-select into study groups, resulting in baseline group differences with respect to characteristics that may influence study outcomes. Thus, differences in outcomes between groups may be due to differences in group characteristics, rather than the effects of the intervention. Selection bias also may be operating for the same reasons when intact or pre-existing groups are used in studies. For example, selection bias likely would be present in a study comparing outcomes of regular patrons of pharmacy A (receiving medication therapy management [MTM] services) versus regular patrons of pharmacy B (receiving usual care) because these intact patient groups may differ with respect to education, health status, or other factors that might affect the outcomes. As controlled experiments use random assignment, selection is not likely to be a threat to internal validity. In both randomized and nonrandomized studies, it is advisable to obtain information on participants' characteristics so differences can be assessed and controlled for in statistical analyses.

History

History refers to the events that occur during the course of the experiment that can affect the outcomes independent of the effects of the intervention. In the MTM study example above, a public service campaign encouraging patients to ask their pharmacists about their medications could be a history threat. Determining whether the study outcomes are due to MTM services, the public service campaign, or both, would be very difficult. However, history is usually not a threat to internal validity in controlled experiments because history events are as likely to affect the intervention as the control group (Polit and Beck 2008). Any differences in outcomes between groups represent effects over and above the history effects; these differences are attributable to the intervention. History becomes a more plausible rival explanation of group differences as the time interval between pre-test and post-test increases (Campbell and Stanley 1963).

Maturation

Polit and Beck (2008) define maturation in a research context as "processes occurring within subjects during the course of the study as a result of the passage of time rather than as a result of a treatment." These changes may occur over long time intervals (participants may become older or more educated) or over short time intervals (participants may become more tired or wound healing may occur naturally). Maturation processes threaten internal validity if they produce outcomes that could be attributed to the intervention. For example, in a study testing the effects of an antibiotic for otitis media, maturation should be considered because some cases resolve naturally without treatment.

Mortality/Attrition

Experimental mortality occurs if participants fail to complete the study. The greatest threat to internal validity occurs when the attrition rates differ between groups, producing groups that are no longer equivalent (Shi 2008). With differential attrition, the advantages of random assignment are lost and conclusions about the intervention may not be valid because between-group differences in the dependent variable may be due to group in-equivalence rather than the intervention. Participants may drop out of a study due to adverse effects of a treatment, dissatisfaction with various aspects of the study, relocation, or other reasons. Higher attrition rates are usually associated with longer data collection periods. An example of attrition threat can be seen in a study evaluating the effectiveness of a diabetes education program in which more participants in the intervention group (compared with the control group receiving usual care) drop out because the program is too demanding.

Testing

Polit and Beck (2008) define testing as "the effects of taking a pre-test on subjects' performance on a posttest." With some types of measures, especially assessments of opinions, attitudes, and knowledge, the process of performing the measurement may change the participants. In some cases, repeated measurements may improve outcomes because each test serves as a practice opportunity; in other cases, administering a questionnaire may sensitize participants to the topic, causing changes in responses on a subsequent administration independent of any intervention. Biophysiologic measures (e.g., serum cholesterol) generally are not subject to testing effects, although there are some exceptions (e.g., having blood pressure taken may increase a patient's blood pressure). Achievement tests and attitudinal measures are more susceptible to this threat. For most measures subject to a testing effect, a longer interval between tests usually results in a smaller testing effect.

Instrumentation

Instrumentation as an internal validity threat refers to changes in measurement instruments or procedures over time, which result in changes in study outcomes independent of the intervention. For biophysiologic measures, improperly calibrated instruments (e.g., blood chemistry analyzers) might yield inconsistent results over time. During patient interviews, more experienced interviewers may elicit different patient responses than less experienced interviewers. In an academic setting, different evaluators may grade a pharmacy student's videotaped counseling session differently, or a single evaluator may grade students' essays more leniently or more harshly while progressing through all of the class essays. Training should be provided to interviewers and other raters to ensure consistency among raters and over time. With appropriate calibration, objective measures (e.g., biophysiologic measures) are less susceptible to instrumentation threats than subjective measures (e.g., observation of behavior or assessment of attitudes).

Statistical regression

Statistical regression (or regression to the mean) refers to the tendency of participants whose scores are extreme on a measure to move closer to the mean on subsequent measures. In other words, participants who scored very high on the pre-test are likely to score lower on the post-test and those who scored very low on the pre-test are likely to score higher on the post-test, because the extreme scores typically contain a larger error of measurement. This becomes an internal validity threat when participants are selected for an intervention based on their extreme scores (e.g., when participants who scored high are selected for a special program as a reward or when participants who scored low are selected for an intervention because they need it more). Regression to the mean should be considered as a rival explanation any time that participants are selected on the basis of extreme scores.

Interactions of internal validity threats

Multiple internal validity threats may be operating simultaneously (Shadish et al. 2002). In addition, internal validity threats may interact with each other. For example, a selection–history interaction might occur in a study of the effects of a MTM program on patient outcomes where the intervention group consists of an intact group of patients from one geographic area and the intact control group is from a different area. These groups might experience different history events that could change study outcomes, making the true effects of the intervention difficult to determine.

Threats to external validity

External validity (generalizability) refers to the extent to which the relationships observed in a study hold true over variations in participants, conditions, settings, interventions, and outcomes (Polit and Beck 2008). For example, the generalizability of the results of a study evaluating treatment algorithms in a Medicaid or Veterans' Administration population is limited to groups with similar characteristics. Also, the generalizability of RCT drug study results to actual use settings is limited. Campbell and Stanley (1963) and Shi (2008) recognize several threats to external validity including the interaction effect of testing, the interaction between selection and the intervention, the reactive effects of experimental arrangements, and multiple treatment interference.

Interaction effect of testing

The interaction effect of testing occurs when taking a pre-test changes the participant's sensitivity or responsiveness to the intervention, thus making the study outcomes unrepresentative of a population who is not pre-tested. For example, this threat would be present in a study of the effects of a continuing education (CE) program on participants' knowledge of new drugs. A pre-test would alert participants to important points in the material and they would be likely to pay more attention when that material is covered, whereas CE participants who are not pre-tested (often the case in CE programs) would not be sensitized to particular aspects of the material. As noted previously, some types of measures (e.g., achievement tests and attitudinal measures) are more susceptible to this threat than others (e.g., biophysiologic measures).

Interaction between selection and the intervention

This threat exists when selection results in participants who are not representative of the population of interest. In this situation, the study participants may respond differently to the intervention than those to whom the researcher wants to generalize. For example, clinical drug trials usually exclude pregnant women and children as participants; however, as a result, the findings of these trials are not generalizable to pregnant women or children and medical professionals are left with a lack of information about use of medications in these groups.

Reactive effects of experimental arrangements

Reactive effects threaten external validity because the artificial environment of the experiment and the participants' knowledge that they are taking part in a study may influence their outcomes. The presence of observers, instruments, or the research environment can have effects that would not be seen when people experience the intervention outside a research study. A famous example of this threat is the Hawthorne effect, which was observed in a study to determine the effects of changes in various environmental conditions (e.g., lighting and working hours) on worker productivity (Kerlinger and Lee 2000). Productivity increased regardless of the change that was implemented because workers knew that they were being observed as part of a study.

Multiple treatment interference

Multiple treatment interference occurs when the same participants experience multiple interventions. The effects of prior interventions tend to persist through subsequent interventions, making it difficult to determine the extent to which any particular intervention affected the outcomes. In this situation, the study results can be generalized only to participants who have experienced the same multiple interventions in the same order. For example, in crossover clinical drug trials, participants in group 1 will receive drug A for a defined period followed by drug B for a defined period, whereas participants in group 2 will receive drug B followed by drug A. In these studies, this threat is minimized by including some interval (a "washout" period) between drug treatment periods.

Validity tradeoffs and priorities

Although researchers attempt to design studies that are strong in all respects, tradeoffs are inevitable. Random assignment and other experimental controls may enhance internal validity but diminish external validity because those experimental conditions differ from natural conditions. For example, taking steps to enhance adherence to a medication regimen in a clinical drug trial would reduce generalizability of the study findings to patients' actual use settings where adherence is likely to be lower. Researchers must establish priorities among the different types of validity based on the research context and purpose, then use research designs, measures, data collection techniques, and statistical analyses that allow them to achieve maximum validity given the established priorities. As a simple example, internal validity is given a higher priority in a clinical drug trial with randomization, double blinding, and controlled medication adherence, whereas external validity is given a higher priority in a study evaluating the effects of an MTM program where randomization and blinding are not possible.

Types of experimental designs

Many experimental designs are used in healthcare research; selected designs that are commonly used or that provide good control of threats to validity are discussed here. The notation used in the figures representing selected experimental designs is based on that used by Campbell and Stanley (1963) in their classic monograph:

- R indicates random assignment of participants to intervention or control groups
- O represents an observation or measurement taken (strictly speaking, O refers to data collected on dependent variables; however, data on independent variables may also be collected at these times)
- X represents the treatment or intervention
- Temporal order is indicated by moving from left to right in the diagrams, and the vertical alignment of Os across groups indicates that these measures are taken at the same time.

Along with descriptions of these designs, particular advantages and disadvantages related to the threats to internal and external validity are noted for each design.

Randomized post-test-only control group design

The simplest randomized experimental design is the randomized post-test-only control group design (Figure 5.1) because participants are randomly assigned to the intervention and control groups, and data are collected only once during the post-intervention phase. Random assignment should produce intervention and control groups that do not differ with respect to characteristics related to the outcome variables of interest. Campbell and Stanley (1963) and Kerlinger and Lee (2000) note that this design generally controls for threats to internal validity. If testing is likely to be a threat to internal validity or to external validity through the interaction of testing and X, this design should be considered because its lack of pre-test measures eliminates the testing effect. However, the design does not allow for the explicit evaluation of pre-to-post changes in the dependent variables (although the control group post-test measure may serve as a proxy for the pre-test measures). This design can be expanded to more than two groups.

| Intervention group | R | X_A | O_1 |
| Control group | R | | O_2 |

Figure 5.1 Randomized post-test-only control group design.

Randomized pre-test–post-test control group design

The randomized pre-test–post-test control group design (Figure 5.2) builds on the randomized post-test-only control group design by adding

Intervention group	R	O_1	X_A	O_2
Control group	R	O_3		O_4

Figure 5.2 Randomized pre-test–post-test control group design.

pre-intervention (baseline) measures that allow for explicit evaluation of pre-to-post changes, and the use of certain statistical analyses that can take advantage of available baseline data (e.g., analysis of covariance). This general design is the most widely used experimental design in drug trials and other clinical research. As with the randomized post-test-only control group design, Campbell and Stanley (1963) and Kerlinger and Lee (2000) note that this design generally controls for threats to internal validity because any existing threats are expected to affect both groups equally. For example, the testing effect may exist, but it should be present in both the intervention and the control groups. Therefore, any changes in the intervention group beyond those in the control group would be attributable to the intervention. However, the design does not control for external validity threats including the interaction of testing and X, which can arise because the pre-test may sensitize participants to the intervention, producing results that would not be seen in a non-pre-tested situation. This design also can be expanded to more than two groups.

Solomon four-group design

The Solomon four-group design (Figure 5.3) consists of four randomly assigned groups with two intervention groups (one pre-tested) and two control groups (one pre-tested). As this design combines the randomized pre-test–post-test control group design and the randomized post-test-only control group design, it possesses the validity control characteristics of both. Thus, it provides effective control for internal validity threats and the external validity threat of the interaction of testing and X (Campbell and Stanley 1963; Kerlinger and Lee 2000). Campbell and Stanley (1963) note that internal validity is especially strong because the effect of X is shown in four different comparisons: $O_2 > O_1$, $O_2 > O_4$, $O_5 > O_6$, and $O_5 > O_3$. If these comparisons are congruent, the strength of inference is increased.

Intervention group	R	O_1	X_A	O_2
Control group	R	O_3		O_4
Intervention group	R		X_A	O_5
Control group	R			O_6

Figure 5.3 Solomon four-group design.

This design has two apparent weaknesses, however: (1) it is resource intensive because it requires more participants, more measurements, and the administration of the intervention to two groups (although they might be combined for the intervention); and (2) statistical analyses are more complicated than for the previous designs because of the unbalanced measures due to the lack of pre-tests for two groups (Kerlinger and Lee 2000). Nevertheless, this design allows for the analysis of the main effects of the intervention, pre-testing, and the interaction of pre-testing and X. Although it can be a very strong design, it is not often used because of its resource needs.

Factorial designs

Factorial designs are used to study the effects of at least two categorical independent variables of interest (each with at least two levels) on an outcome variable (Shadish et al. 2002). Figure 5.4 shows a randomized pre-test–post-test factorial design using two intervention variables (A and B), each with two levels (denoted by subscripts 1 and 2). This design uses four groups to investigate the effects of interventions A and B, with each group receiving a unique combination of the levels of A and B as the intervention. For example, a study of the effects of weight loss programs (diet and exercise [DE] vs diet, exercise, and appetite suppressant [DEAS]) and personal trainers (present [PT-Y] vs absent [PT-N]) on weight (O) would use four groups: (1) DE/PT-Y; (2) DE/PT-N; (3) DEAS/PT-Y; and (4) DEAS/PT-N.

Intervention group	R	O_1	$X_{A_1B_1}$	O_2
Intervention group	R	O_3	$X_{A_1B_2}$	O_4
Intervention group	R	O_5	$X_{A_2B_1}$	O_6
Intervention group	R	O_7	$X_{A_2B_2}$	O_8

Figure 5.4   A randomized pre-test–post-test factorial design.

This design allows researchers to study the combination effects of weight loss programs and personal trainers, and the interaction effect of those two interventions. Control of internal validity threats is enhanced if this design includes a true control group. As in the randomized pre-test–post-test control group design, external validity threats may not be controlled.

Randomized crossover designs

In the designs described so far, each group of participants has received only one treatment. In crossover designs, each group of participants receives multiple treatments. For example, in a crossover design with two interventions, all

Intervention group	R	O_1	X_A	O_2	X_B	O_3
Intervention group	R	O_4	X_B	O_5	X_A	O_6

Figure 5.5 A randomized crossover design.

participants receive both interventions. The participants are randomly assigned to two groups. In one group, they receive intervention A first followed by a post-test, then intervention B followed by a post-test. In the other group, they receive intervention B first followed by a post-test, then intervention A followed by a post-test (Figure 5.5). The design may be extended to more than two interventions and groups.

Advantages of this design include reduced confounding because each participant serves as his or her own control, and increased statistical power so that fewer participants are needed (Grady et al. 2001). However, the duration of the study is increased because the interventions are administered in sequence. In addition, the carryover effect must be considered; if the effects of the first intervention persist into the second intervention, confounding occurs. To avoid this, researchers introduce "washout" periods between interventions so that the effects of the first intervention can dissipate before the second intervention is implemented. Crossover studies are best suited when carryover effects do not persist for long periods and when the number of study participants is limited.

Randomized repeated measures designs

Randomized repeated measures designs usually have one pre-test measure and several post-test measures at defined time intervals for the intervention and control groups (Shadish et al. 2002). For example, a study to determine the effects of a diabetes education program on medication adherence and HbA1c levels may take a pre-implementation baseline measure followed by 3-month, 6-month, 9-month, and 12-month post-implementation measures (Figure 5.6). Participants randomly assigned to the intervention group would take part in the diabetes education program whereas control group participants would receive usual care. This design can be viewed as adding additional post-test measures to the randomized pre-test–post-test control group design. Therefore, it also controls threats to internal validity

Intervention group	R	O_1	X_A	O_2	O_3	O_4	O_5
Control group	R	O_6		O_7	O_8	O_9	O_{10}

Figure 5.6 A randomized repeated measures design.

effectively and can show changes in the dependent variables over time, which might not be seen with a single post-test measure.

With regard to external validity, the interaction of testing and X could be a threat. In addition, results from this design may not be generalizable to situations where multiple observations are not done. Attrition may become a problem with any longitudinal design, especially with longer observation periods; it becomes a threat to internal validity if it occurs differently between the intervention and control groups. Statistical analyses such as repeated measures analysis of variance can be used to determine both between-group and within-participant differences over time.

Other experimental design considerations

Many factors must be considered in choosing the appropriate study design and developing research methods that are appropriate for a given study. In addition to the internal and external validity considerations noted above, other important factors include ethical issues and resources.

Ethical issues

Ethical considerations in any research endeavor are of prime importance, particularly in studies involving human participants or animals. Studies involving humans must be approved by an institutional review board (IRB) to ensure that the potential for harm to participants (physical, psychological, and economic) is minimized and that participants are made aware of the nature of the research, potential risks and benefits, and their rights. The National Institutes of Health and other organizations have their own guidelines and requirements regarding research involving human participants and training in research ethics.

Resource considerations

The resources needed for a study include participants or other data sources, facilities, supplies, equipment, personnel (expertise), time, and funding. Resource considerations affect decisions about all aspects of a research study, including the study design that can be used, the number of participants, the nature of the intervention, the nature and number of observations, and the number and type of researchers that can be involved (which determines the available expertise). As available resources are usually limited, researchers must use them efficiently in order to achieve study goals.

Strengths and limitations of experimental designs

Controlled experiments are regarded as the "gold standard" for scientific research because they can provide strong evidence regarding causal relationships. Through manipulation of the intervention, random assignment to groups, and comparisons of between-group and pre–post measures, these designs can attain a high degree of internal validity, allowing researchers to rule out rival hypotheses in causal relationships.

Despite this major strength, controlled experiments have several important limitations. First, they are not appropriate for all settings and situations; in some cases, quasi-experimental or nonexperimental designs are preferred or more feasible, especially in health services, social science, and policy research. Second, controlled experiments are often conducted in artificial environments which may affect participants' responses and produce results that are not generalizable to natural or actual use settings. Third, consistency of the intervention and the measurements is difficult to maintain, particularly over long periods (Polit and Beck 2008). Fourth, Shadish et al. (2002) and Shi (2008) note that controlled experiments are usually limited with respect to studying only a few of the potential causal factors that affect the outcome of interest in complex ways. Finally, although controlled trials may provide strong evidence regarding causal relationships, they may not answer the questions of why or how that relationship exists. For example, a clinical drug trial may show that an antibiotic is effective against a particular organism, but the mechanism of action may not be revealed.

Summary and conclusions

Researchers, policy makers, and other decision makers are often interested in causal relationships. In studying causal relationships, researchers must consider many factors in selecting an appropriate study design (from many available designs), including how well the design controls for threats to internal and external validity.

Experimental studies, controlled experiments in particular, are best for studying causal relationships. Important characteristics of controlled experiments are manipulation of the intervention, random assignment to intervention and control groups, and comparisons of outcomes (between-group and pre–post comparisons). These characteristics contribute to strong internal validity, but controlled experiments often lack external validity. They also are not always feasible and have several other limitations. Nevertheless, controlled experiments are powerful tools that researchers use to explore causal relationships between various interventions and outcomes in medical research and other areas.

Review questions/topics

1 What criteria must be satisfied to establish a causal relationship?
2 What are the three major design characteristics of controlled experiments and how do they contribute to study validity?
3 Describe important differences of laboratory experiments, field experiments, natural experiments, and simulation experiments.
4 Define the threats to internal and external validity, and describe how each threat may affect study validity.
5 Describe the types of experimental designs and how each design controls internal and external validity threats.
6 What are the major strengths and limitations of controlled experiments?

References

Campbell DT, Stanley JC (1963). *Experimental and Quasi-Experimental Designs for Research*. Boston, MA: Houghton Mifflin Co.

Cummings SR, Grady D, Hulley SB (2001). Designing an experiment: Clinical trials I. In: Hulley SB, Cummings SR, Browner WS, *et al.* (eds), *Designing Clinical Research*, 2nd edn. Philadelphia: Lippincott Williams & Wilkins, 143–55.

Grady D, Cummings SR, Hulley SB (2001). Designing an experiment: Clinical trials II. In: Hulley SB, Cummings SR, Browner WS, *et al.* (eds), *Designing Clinical Research*, 2nd edn. Philadelphia: Lippincott Williams & Wilkins, 157–74.

Kerlinger FN, Lee HB (2000). *Foundations of Behavioral Research*, 4th edn. Belmont, CA: Cengage Learning.

Polgar S, Thomas SA (1991). *Introduction to Research in the Health Sciences*. Melbourne: Churchill Livingstone.

Polit DF, Beck CT (2008). *Nursing Research: Generating and assessing evidence for nursing practice*, 8th edn. Philadelphia: Lippincott Williams & Wilkins.

Rand Health (2008). Rand Health Insurance Experiment (HIE). Online. Available at: www.rand.org/health/projects/hie (accessed 26 August, 2009).

Saks M, Alsop J (2007). *Researching Health: Qualitative, quantitative and mixed methods*. Los Angeles, CA: Sage Publications.

Shadish WR, Cook TD, Campbell DT (2002). *Experimental and Quasi-experimental Designs for Generalized Causal Inference*. Boston, MA: Houghton Mifflin Co.

Shi L (2008). *Health Services Research Methods*, 2nd edn. Clifton Park, NY: Delmar Learning.

Soumerai SB, Ross-Degnan D, Avorn J, McLaughlin TJ, Choodnovkiy I (1991). Effects of Medicaid drug-payment limits on admission to hospitals and nursing homes. *N Engl J Med* 325: 1072–7.

St Charles ME, Sadri H, Minshall ME, Tunis SL (2009). Health economic comparison between continuous subcutaneous insulin infusion and multiple daily injections of insulin for the treatment of adult type 1 diabetes in Canada. *Clin Ther* 31: 657–67.

Online resources

National Cancer Institute (1998). *Investigator's Handbook: A manual for participants in clinical trials of investigational agents*. Available at: http://ctep.cancer.gov/investigatorResources/docs/hndbk.pdf.

National Institutes of Health. "Frequently Asked Questions for the Requirement for Education on the Protection of Human Subjects." Available at: http://grants.nih.gov/grants/policy/hs_educ_faq.htm.

National Institutes of Health Office of Human Subjects Research. Available at: http://ohsr.od.nih.gov.

6

Nonexperimental research

Michael L Johnson

Chapter objectives

- To examine the role of nonexperimental research
- To discuss common nonexperimental research designs
- To present approaches to control confounding
- To identify strengths and weaknesses of nonexperimental designs

Introduction

In Chapter 5, experimental designs were presented involving manipulation and control of the experimental conditions by the investigator and random allocation of the study participants to the experimental conditions. In non-experimental, or observational, studies, the investigator does not have control over these features of the study. The study conditions themselves may or may not be chosen by the investigator, but the allocation of participants to experimental conditions or treatment groups is not implemented by the investigator. For example, an investigator may wish to study the effect of antihypertensive agents on the risk of cardiovascular disease. The investigator conducts a literature review and determines that the most important outcomes to study are stroke, myocardial infarction, and death. In a sense, the investigator has little choice over the conditions to study: they are just "there" as the disease conditions that have been discovered to be the most important by scientists and clinicians. It would make no sense for the investigator to choose some other condition, such as the effect of antihypertensive agents on IQ. The investigator may also have little choice over the hospitals or clinics from which to select patients to study. When the investigator enrolls study participants, it is found that they are already taking, or not taking, an array of possible antihypertensive agents. The investigator does not assign patients to an angiotensin-converting enzyme (ACE) inhibitor or a diuretic or other treatment. The patients are observed to be taking whatever treatments they are prescribed by their physicians. If the investigator wanted to assign patients to take certain antihypertensive agents, perhaps this could be done, but then this would become an experimental study.

The lack of control over the allocation of the study participants to treatment conditions is a serious challenge to investigators conducting nonexperimental research. In a very real sense, control is the issue! Why would an investigator choose to give up this control? There are many instances when experimental or randomized intervention studies are simply impractical or indeed unethical in the study of human disease epidemiology. This chapter discusses the role of nonexperimental research in pharmaceutical practice and policy sciences, and examines study designs appropriate to nonexperimental research, with a focus on approaches to mitigate confounding and bias. Strengths and limitations of nonexperimental research designs are discussed to place these studies in proper context.

Role of nonexperimental research

Health insurers, physicians, and patients worldwide need information on the effectiveness and safety of prescription drugs in routine care. Nonexperimental studies may be able to address some of the limitations of experimental studies, and complement or otherwise add to existing knowledge of drug safety and effectiveness in ways that controlled trials cannot.

Suppose that an investigator wishes to study the risk of liver cancer in patients taking statins. Liver cancer is a very rare disease, and would require enrollment of a very large number of patients, requiring years of follow-up, before enough patients would develop the outcome for a valid study (El-Serag et al. 2009). An experimental study would require that some patients be given statins and others not be given statins. In patients with hyperlipidemia, it may be unethical to randomize patients not to receive lipid-lowering therapy. In patients without hyperlipidemia, it might be unethical to enroll patients to receive statins. How would this clinical question be studied?

Currently, there is reluctance by many health policy decision makers to use observational data – especially data from retrospective analysis of large datasets – to inform their deliberations. Many decision makers are uncertain about the validity of results derived from observational studies, primarily due to concerns about confounding and selection bias.

Although randomized clinical trials (RCTs) are the gold standard to determine a drug's efficacy against placebo, it is well recognized that results of such studies may not accurately reflect the effectiveness of therapies delivered in typical practice to patients who are more representative of larger populations (Concato et al. 2000; Concato 2004; Avorn 2007). Observational studies can be designed and conducted with rigorous techniques that improve internal validity, with advantages that address limitations of RCTs. To make these

studies valid and useful to policy makers, investigators must pay speci
attention to confounding and bias.

Epidemiologic framework

One way to place nonexperimental research into perspective is to consider the
epidemiologic framework of studies of disease in humans. Epidemiology is the
branch of science that investigates the causes and control of epidemics in
populations. Epidemiology derives from the Greek *epi* which refers to
"around" or "among" and *demios* meaning "the people." The science originally dealt with infectious diseases, diseases that "go around or among the
people," but now includes chronic disease. The epidemiologic framework
then considers the causes of disease in terms of exposure to infectious or
environmental agents, or other risk factors for the disease. The framework
can be expanded to include anything that can be considered as an "exposure,"
such as treatment with a certain drug, as well as any kind of "outcome" such
as morbidity or mortality. Other terms for exposure could be risk factor,
patient factor, independent variable, or anything that is defined by an investigator as a factor that may lead to a disease or outcome of interest. Other
terms for outcome could be case, event, dependent variable, or anything that
is defined by an investigator as an outcome. The epidemiologic framework
(Figure 6.1) is a flexible and powerful way to conceptualize research into
causes and effects of a wide range of constructs in people.

Studying the treatment–outcome relationship

In RCTs, identifying and measuring exposure is done with a great deal of
accuracy and precision. For example, in clinical trial evaluation of drug
treatment, it is known not only who has received the active drug, but also
the degree of exposure – dose, duration, and compliance with therapy.

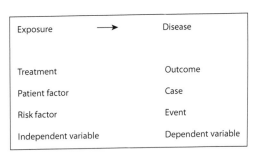

Figure 6.1 The epidemiologic framework of the relationship of exposure and disease can be
expanded to include other concepts such as treatment and outcome, risk factors, and events.

1

utcomes, or measures of effectiveness, are measured with a great
uracy and precision. Various devices and laboratory tests are used
e and record both surrogate (blood pressure, cholesterol levels,
ging) and final endpoints (e.g. myocardial infarction, stroke, and
h). This same level of precision is often not universally available in
onal studies.

of nonexperimental research

key study designs used in observational research are outlined below.
views of observational study designs are available (see Lu 2009).
nes have also recently been proposed on the reporting of observational
s, specifically as it relates to cross-sectional, cohort, and case–control
es (von Elm et al. 2008), as well as issues surrounding the design and
lysis of these studies for comparative effectiveness (Berger et al. 2009;
x et al. 2009; Johnson et al. 2009).

Epidemiologic designs

Cohort designs

In a cohort study, groups of patients (i.e., cohorts) exposed, or not exposed, to
drug therapies are followed over time to compare rates of one or more outcomes of interest between the study cohorts. Patients are enrolled at a baseline
starting point, and their exposure to the risk factor of interest is measured.
They are then followed forward in time to determine the outcome of interest.
These types of studies are called prospective cohort studies. Temporal relationships between exposure and outcome can be well characterized in a
cohort study and both relative and absolute risks can be reported directly
with the use of this design. Consequently, this design may be of particular
interest for research questions requiring absolute risk estimates and where the
temporal nature of associations is important to characterize. The design is
suitable for various types of exposure even if the exposures are rare, and can
be conducted retrospectively, by analyzing existing exposure–disease information collected at some point in the past.

Case–control designs

Case–control designs involve the identification of individuals who experience
an outcome of interest (i.e., cases) and those who do not (i.e., controls).
Patients are then examined backward in time to determine their exposure
status to a risk factor of interest. Exposure to a drug of interest in a period
before the designation of case or control status is then compared between
cases and controls. This design has historically been used when the outcome of
interest is rare, maximizing the capture of such precious outcomes. Analysis of

case–control designs typically provides estimates of relative risk but does not directly provide absolute risk estimates.

It is important to remember that the exposure–disease relationship is of primary interest. Cohort studies allow the investigator to study this relationship in a temporally forward motion, from exposure to disease, in a way that makes sense epidemiologically: the exposure occurs first, and then causes associated with subsequent disease. The case–control study is a clever way to examine the same relationship, just in reverse. The outcome or disease is identified first, and then medical records or exposure history is examined to determine if exposed individuals were more or less likely to have the disease. It is just a different way to look at the same scientific question (exposure → disease relationship).

Nested case–control

In the nested case–control study design, a case–control study is conducted within the sample obtained from a cohort study. This study design is a hybrid of both the cohort and the case–control studies. In the study by El-Serag and colleagues (2009), the investigators wished to study whether the use of statin drugs affected the risk of developing liver cancer. The investigators chose to study this relationship within a cohort of patients identified with diabetes. After first excluding any prevalent or pre-existing cases of liver cancer, all new incident cases of liver cancer were identified. Once these cases were identified, a sample of patients without liver cancer was obtained, and matched on several factors. The medical history was examined for exposure to use of statins, to compare rates of statin use among those with and those without the liver cancer. It was found that patients who had liver cancer were less likely to be using statins. The investigators then infer a possible protective effect of statin use on liver cancer in patients with diabetes.

The main feature of a nested case–control study that distinguishes it from a case–control study is the well-defined nature of the study sample. This study design maintains all the advantages and disadvantages of both the cohort and the case–control study. There must still be adequate sample size and follow-up time for the cohort to develop the outcome if it is rare, but rare outcomes can then be studied as in case–control studies, and with the added benefit of a much clearer definition of the population being studied.

Case–crossover designs

A primary challenge of cohort and case–control studies is the selection of comparable comparison groups. In case–crossover studies only those individuals who experience the outcome of interest (i.e., cases) and were exposed to a treatment of interest within a certain time before the outcome date are included. Individuals serve as their own controls. Exposure to the treatment of interest in the period immediately before the outcome is compared with exposure prevalence in a more distant period to the event of interest in the same individual. Exposure prevalences are then compared between more recent and distant exposure windows to arrive at a risk ratio. Case–crossover designs are ideally suited for transient exposures that result in acute events but require sufficient numbers of patients who both have an event and are exposed to the drug of interest in either the recent or more distant exposure windows. This design may be particularly attractive for research questions involving the comparison of groups that are extremely different in their clinical profiles (i.e., where major selection bias may exist) and involve transient exposures and immediate outcomes.

Case–time–control designs

A limitation of case–crossover designs is temporal confounding, where the prevalence of treatment exposure is higher when closer to the event date than farther from the event date simply because of naturally increasing treatment uptake over time, rather than a truly causal relationship. To circumvent this issue, case–time–control studies create a control group of individuals who do not experience the event of interest and analyze the group in a manner similar to the cases to estimate the "natural" increase in treatment exposure prevalence over time – the exposure prevalence in the exposure window closer to the event date is compared with the exposure prevalence in the exposure window in a more distant period to arrive at a risk ratio among controls. The "case" risk ratio is then divided by the "control" risk ratio to arrive at an overall risk ratio. This design also requires sufficient numbers of patients who have both an event and exposure to the treatment of interest in either of the predefined exposure windows. Issues of selection bias in comparing cases with controls may still be problematic (Greenland 1996).

Other designs

Cross-sectional designs

The cross-sectional study examines a "snapshot" of data and either describes the data available in that snapshot or attempts to make correlations between variables available in the dataset. Although this study design can provide some valuable information, it is limited by its inability to characterize temporality – it is often uncertain whether the exposure preceded the outcome of interest or vice versa. In research questions where temporality of exposure and outcome is important, alternate designs should be selected.

Ecological designs

Studies that examine the association of factors at group or aggregate levels are ecological or correlational studies. Measures of exposure and outcome are made and examined at the group level. For example, the rates of ACE inhibitor use for all the patients at hospitals in a certain county could be measured, and the rates of heart attack in those patients examined. It may be that, as the rates of ACE inhibitor use increase across hospitals, so do the rates of heart attack (positive correlation). A major drawback of ecological studies is to infer individual effects from the group associations, called the ecological fallacy. To illustrate, suppose that, in the hospital with the lowest rate of both ACE inhibitors and heart attack, the patients who took the ACE inhibitors had fewer attacks than the patients who did not (negative correlation). Suppose that this pattern were true across all hospitals. It would be wrong to infer the relationship of ACE inhibitors and heart attacks in individual patients from the group study. It is therefore of the utmost importance in ecological studies to be cautious in making inferences at the individual level from data collected at the aggregate level.

Pre- and post-designs

Pre–post study designs are weak designs but are commonly used. They occur when a group is measured before some intervention and then after the intervention, and the resulting difference in outcome is measured and attributed to the effect of the intervention. Such designs are fraught with threats to validity. Many other factors may have contributed to any changes in the outcome variable, such as other interventions that have may have occurred simultaneously. These studies are also very susceptible to a pernicious statistical problem known as regression to the mean which can plague even the most experienced investigator. Observations at one point, say pre-intervention, tend to approach the mean for a given measure on subsequent observations. In a one-group pre–post experiment, it is entirely likely that, if a group has relatively high cholesterol before an intervention, that group may be expected to have lower cholesterol readings in subsequent measures, due to any kind of reason, such as diet or exercise, which might lower cholesterol, or natural variation in a person's cholesterol readings.

To improve interpretations of results from pre–post studies, it is wise to include a control group that was not exposed to the treatment or intervention, and take measures both before and after on this group as well. The variation in the differences between the groups can then be examined to determine if there is any "real" effect of the intervention, because the control group would also have all the effects of other possible factors, including regression to the mean. This study design cannot, however, control for unknown factors that might result in differences in trends occurring before the intervention between the study and control groups.

Interrupted time series designs

Interrupted time series analysis involves cross-sections of data over time both before and after an event of interest. Actual trends in exposures or outcomes after an event of interest are then compared with the expected trends based on patterns of historical data before the event of interest. For example, in assessing the impact of a drug policy on drug utilization, historical trends would be used to establish an expected drug utilization rate in the absence of the policy change (Mamdani et al. 2007). This expected drug utilization rate would then be compared with observed rates occurring after implementation of the drug policy using advanced statistical approaches. The benefit of conducting a time series analysis is the minimization of the problematic selection bias. Challenges, however, include issues related to temporal confounding (i.e., other events that may have occurred simultaneously at the time of intervention) and the need to be studying policies with relatively large effects on the outcomes. This design may be particularly relevant for research questions aimed at assessing the impact of policy decisions on drug utilization and immediate outcomes.

Bias and confounding in nonexperimental studies

Bias is any systematic deviation or error in measurement. Confounding is a type of bias where the exposure–disease relationship is distorted by a third factor.

Bias

Perhaps the most important problem in nonexperimental studies is selection bias. Selection bias occurs when there is any differential classification of participants with respect to treatment or outcome. If patients are selected into the study differently, based on how they are treated, this leads to bias, and indeed is often the case in observational drug treatment studies. If the determination of outcome is measured differently in patients with the outcome than in patients without the outcome, this leads to bias. Bias can therefore lead to erroneous conclusions about the treatment–outcome relationship. For example, it may be that a patient is determined not to be taking the drug being studied, when in fact the patient is taking that drug. Or, in terms of outcomes, the patient could be classified as having angina when in fact the patient really had anxiety.

A common type of selection bias has been called prevalence bias. Patients included in cohort studies for a given disease condition have varying lengths of time at risk for the disease, and varying lengths of duration of time with the disease. Patients are then treated for the disease, and may have varying lengths of time since starting the treatment. If an investigator begins to study patients at a certain point in time, and wants to follow them and measure the occurrence of an outcome, the investigator must be aware that some patients may have already taken a certain drug for a long period of time, may have already taken that drug and then stopped taking it, or some patients may initiate use of the drug for the first time during the course of follow-up. Suppose an investigator wishes to examine whether patients treated in a managed care population are attaining target goals for cholesterol and blood pressure (Johnson et al. 2006). If patients who have been taking lipid-lowering agents and antihypertensive agents for a long time are included with patients who are newly prescribed these drugs, results could be biased. Depending on the research questions being investigated, a common way to control for prevalence bias is to include new users of drugs, after first excluding prevalent users (Ray 2003).

Measurement of drug exposure in observational research is susceptible to a special kind of bias called classification bias – identifying participants as being exposed to a drug when they are not exposed, or not exposed when they are. Classification bias is further categorized as differential or nondifferential and unidirectional or bidirectional. Nondifferential misclassification occurs when the likelihood of misclassification is the same across the exposed or outcome groups. For example, classification bias of exposure for a low-cost medication using prescription claims data would be equally likely regardless of outcome. However, differential misclassification is present when the likelihood of misclassification is different between exposed and outcome groups. An example of differential misclassification for drug exposure is when those who are exposed have a lower likelihood of outcome misclassification because to receive medication they have to enter the healthcare system, which increases their likelihood of recording a diagnosis. Those not exposed are much more likely to be misclassified as not having the disease, which is an artifact of not entering the healthcare system.

Confounding

Confounding is classically defined as a bias that distorts the exposure–disease or exposure–outcome relationship (Miettinen 1974). Frequently used definitions of confounding and standard textbook methods to control for confounding state that a confounder is an independent (causal) risk factor for the outcome of interest that is also associated with the exposure of interest in the population, but is not an intermediate step in the causal pathway between the exposure and the outcome (Grayson 1987; Weinberg 1993).

Confounding by indication for treatment

A common and pernicious problem endemic to pharmacoepidemiologic studies is confounding by indication for treatment. For example, when the choice of therapy is affected by the severity of illness, and physicians prescribe one therapy over another depending on the severity and the perceived effectiveness of one drug, compared with another, for patients with differing severity levels, then confounding by indication for treatment occurs (assuming that the severity of disease is also a risk factor for the outcome of interest). In this case, apparent treatment effects are confounded, i.e., they are not causal but they may actually be caused by the severity of illness that led to patients being prescribed a given treatment.

Measured vs unmeasured confounding

Confounders may be measured or unmeasured. Secondary databases of a variety of sources may contain a wide and rich variety of information that can be used to measure an array of potentially confounding factors. However, even the most detailed and complete data sources may fail to include information on potential confounding factors, and these remain unmeasured and hence uncontrolled in a given study, leading to residual confounding. Methods to address both measured and unmeasured (residual) confounding factors have been developed to address these concerns (Johnson et al. 2009).

Time-dependent confounding

The more complicated (but probably not less common) case of time-dependent confounding refers to variables that simultaneously act as confounders and intermediate steps, i.e., confounders and risk factors of interest mutually affect each other (Figure 6.2). Confounding by indication may take the form of time-dependent confounding. An example is the

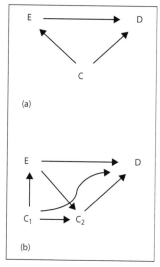

Figure 6.2 Simple diagram showing (a) point- or time-independent confounding and (b) time-changing or time-dependent confounding. E, exposure or treatment of interest; C, confounder; D, disease (or any other outcome). In (a) confounding occurs at a point in time (similar to a point estimate of a parameter); in (b) the time-independent confounder can change (now time-dependent) and is measured at two points in time, C_1 and C_2. (Adapted from Cox et al. 2009.)

effect of aspirin use (treatment) on risk of myocardial infarction (MI) and cardiac death (outcome). Prior MI is a confounder for the effect of aspirin use on risk of cardiac death, because prior MI is a cause of subsequent aspirin use, and also a causal risk factor for subsequent cardiac death. However, prior aspirin use also causally prevents prior MI. Therefore, prior MI simultaneously acts as a confounder (causing aspirin use) and an intermediate step (being affected by aspirin use), and hence is a time-dependent confounder affected by previous treatment.

Another example of time-dependent confounding by treatment is antiviral treatment of HIV infection, where treatment or dose may depend on the CD4 count, and this dependency may continue over the course of the disease (Hernan et al. 2000).

Approaches to mitigate confounding and bias in design

The main tools to address confounding and bias in the design stage are restriction of study patients into the sample, and extremely careful attention to data collection. Bias generally creeps into studies during the conduct of the study, by either misclassification of study participants or differential selection of study participants. Bias also occurs with differential measurement of variables, most commonly in terms of missing data. It is very difficult to adjust for bias in the analysis stage of studies, so it is extremely important to address it as early as possible in the design stages.

Perhaps the most common standard technique to address selection bias, which also addresses some of the confounding factors, is restriction of study participants into the sample by inclusion and exclusion criteria. Rather than just study "everyone," investigators even in observational studies must make a range of decisions on whom to include and whom to exclude. Restricting study cohorts to patients who are homogeneous with regard to their indication for the study drug will lead to more balance of patient predictors of the study outcome among exposure groups and thus will reduce confounding. This will not, however, necessarily eliminate confounding, particularly when variables that influence prescribing decisions are not available in the data. Restricting study cohorts can also increase the likelihood that all included participants will have a similar response to therapy and therefore reduce the likelihood of effect modification, or interaction. RCTs commonly restrict their study population to patients with a presumed indication for the study drug. In observational studies, particularly of elderly people with possibly multiple chronic conditions, drugs may be indicated for more than one condition. Schneeweiss (2007) provides a very readable discussion of bias and confounding in nonrandomized studies of effectiveness and safety of medical interventions.

Analytic methods to adjust for confounding factors

Common methods to adjust for confounding factors in the analysis stage include stratification, matching, and regression techniques.

Stratification

Stratified analysis is a fundamental method in observational research that involves placing data into subcategories, called strata, so that each subcategory can be observed separately. Its many uses in observational studies include standardization, control of confounding, subgroup analysis in the presence of effect–measure modification, and addressing selection bias of the type that occurs in matched case–control studies.

Stratification is an intuitive and hands-on method of analysis, results are readily presented and explained, and it does not require restrictive assumptions. As stratified analysis can be applied "hands on," often in a simple spreadsheet program, it allows investigators to get closer to their data than they otherwise could by using more complex methods such as multivariable regression. Its disadvantages include a potential for sparsely populated strata, which reduce precision, and cause a loss of information when continuous variables are split into arbitrarily chosen categories and a tendency to become arduous when the number of strata is large. For more information, see a practical and comprehensive discussion of stratified analysis by Rothman et al. (2008).

Matching

Matching is a form of blocking on factors that are expected to be associated with outcome, but are not the factors to be directly examined in the given study. For example, in a study of cardiovascular disease and risk of 1-year mortality, it may be advisable to control or adjust for the age difference of patients in the study. One way to do this would be to match patients receiving a certain treatment with those who are not, based on their age. For every patient receiving a treatment, another patient of the same age who is not receiving the treatment is selected, thus reducing possible confounding effects of age. Other factors commonly matched in observational studies are ethnicity, sex, and time of exposure. Two important considerations in matched study designs are that an appropriate matched analysis should be conducted and that the effects of these factors cannot then be estimated because they have been equalized as a result of the matching.

Regression

Regression is a powerful analytical technique that can accomplish several goals at once. When more than a few strata are formed for stratified analysis, or when more than a few potential confounding factors need to be adjusted, multiple regressions can be used to determine the unique association between the treatment and the outcome, after simultaneously adjusting for the effects of all the other independent factors included in the regression equation. In a multiple regression equation, the effect of a treatment variable has been independently estimated after estimating the effects of all the other variables in the regression. Another very important use of regression is to use the regression equation to predict study outcomes in other patients. This is the primary use of multiple logistic regression when used for propensity scoring. For example, an investigator may formulate a regression model that predicts the occurrence of heart attacks in patients with high blood pressure, and estimates the effect of an antihypertensive agent. The investigator may then want to apply this predictive model to a different population in order to estimate how many patients treated or not treated with that drug may be expected to have heart attacks, based on the predictive model. The predictive model may also be used to examine determinants of being or not being treated with the drug.

A look at advanced techniques

Propensity score analysis

The propensity score is an increasingly popular technique to address issues of selection bias and confounding by indication, commonly encountered in observational studies estimating treatment effects. The propensity score is defined as the conditional probability of being treated given an individual's covariates (Rosenbaum and Rubin 1983; D'Agostino 1998). The main idea of propensity scoring is quite intuitive. In an RCT, patients are randomly assigned to a treatment group or a comparison group. All the potential confounding factors that could affect the treatment–outcome relationship have been balanced by the randomization between the two groups, so that any differences in outcome are the result of the treatment. In observational studies, the treatment groups are not randomly assigned, and there are a large number of factors that may be associated with whether or not a patient receives a certain treatment. Suppose that the investigator measures all these factors, and conducts a logistic regression model to predict whether or not the patient receives the treatment, using all these factors as predictor variables. The resulting predicted probabilities of receiving treatment are obtained from the regression model. The theory behind propensity scoring confirms that, for two patients with the same predicted probability of receiving the treatment, of whom one actually received the treatment and one did not, the two patients are essentially the same as two patients who were or were not randomized to receive treatment. Patients who were or were not treated, and are then matched on their propensity scores, thus form pseudo-randomized groups of patients who can then be studied for differences in treatment effects on outcomes.

Instrumental variables analysis

Another approach to try to adjust for confounding factors is the instrument variable approach which is very common in econometric literature. It is impossible to measure all the reasons that doctors prescribe certain medications, but a physician's prescribing preference is known to be an important determinant of differential treatment selection. If the prescribing preferences cannot be measured, they cannot be controlled, and the results of treatment effects in an observational study would be biased. Briefly, an instrumental variable is sought that is associated with the confounding factors but not with the outcome. If this variable can be found and included in a regression model, the effects of the confounding factors can be adjusted for even if they are not directly measured. An example is a study by Brookhart et al. (2006) in which the investigators measured the most recent prescription by a physician as a measure of his or her prescribing preference. The prescription given to a previous patient could not possibly affect the outcome of a subsequent patient, but may be a strong indicator of the same physician preferences which may have led to the drug chosen being prescribed to the subsequent patient.

Strengths and limitations of nonexperimental studies

There are many strengths in nonexperimental study designs, as well as many limitations. Primary strengths compared with those of experimental studies include increased external validity, understanding real-world effectiveness of drug treatments, the ability to capture long-term exposures, and detection of unintended risks and benefits, as well as hypothesis generation which may be tested later in experimental studies.

Although experimental studies have strong internal validity, they have weaker external validity compared with observational studies. A study of all the patients with diabetes in a managed care plan that serves a spectrum of millions of patients across the entire USA has broad applicability to millions of people. A closely related advantage is the estimate of treatment effectiveness in so-called "real-world" settings. It is extremely difficult for RCTs to examine treatment effects in everyday use. Patients enrolled in well-designed controlled trials are followed, observed, and measured in ways that are often quite different from patients treated in normal care settings. Effects estimated in observational studies, therefore, even if not as precisely measured as in experiments, at least reflect the treatment effects that people, doctors, and patients in real life will likely encounter. Longer-term exposures can be examined, which may lead to insights with regard to drug safety that cannot be determined from short-term trials. Observational studies also confer the ability to study treatment effects that have perhaps not been expected. For example, it was a series of epidemiologic observational studies showing increased risk of rofecoxib on MI and stroke that eventually led to its recall by the US Food and Drug Administration. Finally, observational studies may elicit findings that are unusual or interesting and can lead to future studies to confirm or deny new treatment effects. Such hypothesis-generating studies are an additional benefit from exploratory observational research.

Limitations of nonexperimental studies revolve primarily around weaker internal validity and interpretation of causal effects. The primary strength of experimental studies is their very strong internal validity by design. The beauty of randomization confers an extremely powerful ability to draw causal inference of treatment effects from differences in outcomes. As discussed, nonexperimental studies do not inherently have this feature. Observational studies therefore must take on an array of problems in the design and analysis to improve causal inference and increase internal validity. Scientists do not agree whether such studies are in fact as reliable as experimental studies. Finally, where observational studies may have increased external validity, they may also have too heterogeneous a population to be able to make clear recommendations to specific groups.

Summary and conclusions

Nonexperimental, or observational, studies offer an alternative to experimental study designs that can complement and in some cases address limitations of experimental research. There is a role for observational research in pharmaceutical policy and practice, especially in today's environment where there is often a lack of fundamental evidence upon which to base not only treatment, but also reimbursement, or other policy decisions. Undoubtedly, there is a range of challenges in conducting such studies, so that findings from these studies are meaningful and useful to patients, physicians, pharmacists, and policy makers, as well as other researchers. However, many of these challenges are extremely well understood by researchers in the field, and are readily addressed by basic study design and analysis techniques. Other challenges remain more daunting, but researchers continue to explore approaches and improve upon methods. The promise, and pitfalls, of observational research in pharmaceutical practice and policy arena are only now beginning to be established. Without question, nonexperimental designs in pharmaceutical practice and policy research are only expected to grow in the coming decades.

Review questions/topics

1 What is the primary difference between a nonexperimental study design and an experimental study design? Why would investigators choose such an approach?

2 Name the major epidemiologic study designs, and briefly describe the main design features for each.

3 What are two main challenges or problems that are inherent in observational studies?

4 Give an example of selection bias in a cohort and a case–control study, and explain how these could affect the results.

5 What is the classic epidemiologic definition of confounding? Give an example.

6 What are a few of the most common approaches to adjusting for confounding in the design and analysis stages?

7 Name three strengths and three weaknesses of nonexperimental studies. Do you think that they are as good as experimental studies? Why or why not?

References

Avorn J (2007). In defense of pharmacoepidemiologic studies: embracing the yin and yang of drug research. *N Engl J Med* 357: 2219–21.

Berger M, Mamdani M, Atkins D, Johnson ML (2009). Good research practices for comparative effectiveness research: Defining, reporting and interpreting non-randomized studies of treatment effects using secondary data sources. The ISPOR Good Research Practices for Retrospective Database Analysis Task Force Report – Part I. *Value in Health* 12: 1044–52.

Brookhart MA, Wang P, Solomon DH, *et al.* (2006). Evaluating short-term drug effects using a physician-specific prescribing preference as an instrument variable. *Epidemiology* 17: 268–75.

Concato J (2004). Observational versus experimental studies: what's the evidence for a hierarchy? *NeuroRx* 1: 341–77.

Concato J, Shah N, Horwitz RI (2000). Randomized, controlled trials, observational studies, and the hierarchy of research designs. *N Engl J Med* 342: 1887–92.

Cox E, Martin BC, Van Staa T, Garbe E, Siebert U, Johnson ML (2009). Good research practices for comparative effectiveness research: approaches to mitigate bias and confounding in the design of non-randomized studies of treatment effects using secondary data sources. The ISPOR Good Research Practices for Retrospective Database Analysis Task Force Report – Part II. *Value in Health* 12: 1053–61.

D'Agostino RB Jr (1998). Propensity score methods for bias reduction in the comparison of a treatment to a non-randomized control group. *Stat Med* 17: 2265–81.

El-Serag HB, Johnson ML, Hachem C, Morgan RO (2009). Statins are associated with a reduced risk of hepatocellular carcinoma in a large cohort of patients with diabetes. *Gastroenterology* 36: 1601–8.

Grayson DA (1987). Confounding confounding. *Am J Epidemiol* 126: 546–53.

Greenland S (1996). Confounding and exposure trends in case-crossover and case-time-control designs. *Epidemiology* 7: 231–9.

Hernan MA, Brumback B, Robins JM (2000). Marginal structural models to estimate the causal effect of zidovudine on the survival of HIV-positive men. *Epidemiology* 11: 561–70.

Johnson ML, Pietz K, Battleman D, Beyth RJ (2006). Therapeutic goal attainment in patients with hypertension and dyslipidemia. *Med Care* 44: 39–46.

Johnson ML, Crown W, Martin BC, Dormuth CR, Siebert U (2009). Good research practices for comparative effectiveness research: analytic methods to improve causal inference from non-randomized studies of treatment effects using secondary data sources. The ISPOR Good Research Practices for Retrospective Database Analysis Task Force Report – Part III. *Value in Health* 12: 1062–73.

Lu CY (2009). Observational studies: a review of study designs, challenges and strategies to reduce confounding. *Int J Clin Pract* 63: 691–7.

Mamdani M, McNeely D, Evans G, *et al.* (2007). Impact of a fluoroquinolone restriction policy in an elderly population. *Am J Med* 120: 893–900.

Miettinen OS (1974). Confounding and effect modification. *Am J Epidemiol* 100: 350–3.

Ray WA (2003). Evaluating medication effects outside of clinical trials: new-user designs. *Am J Epidemiol* 158: 915–20.

Rosenbaum PR, Rubin DB (1983). The central role of propensity score in observational studies for causal effects. *Biometrika* 70(1): 41–55.

Rothman KJ, Greenland S, Lash TJ, eds. (2008). In: *Modern Epidemiology*, 3rd edn. Philadelphia: Lippincott Williams & Wilkins, 258–302.

Schneeweiss S (2007). Developments in post-marketing comparative effectiveness research. *Clin Pharmacol Therapeut* 82: 143–56.

Von Elm E, Altman DG, Egger M, *et al.* (2008). The Strengthening the Reporting of Observational Studies in Epidemiology (STROBE) statement: guidelines for reporting observational studies. *J Clin Epidemiol* 61: 344–9.

Weinberg CR (1993). Towards a clearer definition of confounding. *Am J Epidemiol* 137: 1–8.

Online resources

International Society for Pharmacoepidemiology. Available at: https://www.pharmacoepi.org/index.cfm.

International Society for Pharmacoeconomic and Outcomes Research. Available at: http://www.ispor.org.

STROBE Statement: Strengthening the Reporting of Observational Studies in Epidemiology. Available at: www.strobe-statement.org/Checklist.html.

7

Sampling methods

Rajender R Aparasu

Chapter objectives

- To explain the sampling terminology
- To describe the steps in a sampling plan
- To discuss the determinants of sample size
- To estimate sample size for a specific objective

Introduction

Pharmaceutical practice and policy research is often based on a sample rather than the population. Sampling involves selection of a small number of observation units from the population. The population includes all available observation units of interest. Sampling is a scientific process of selecting a subset of observable units to address a research question. Sampling methodology has practical and scientific value in conducting research. The sampling process can be instrumental in developing detailed plans for the research. It also helps in planning for resources needed to implement a research study. It provides a cost-effective way to conduct the study without the need to collect the data from all observable units in the population. An appropriately selected sample will provide findings that have internal and external validity. Several factors influence the sampling plan such as research question, research design, data collection methods, and data analysis. This chapter explains the sampling terminology that is important in implementing a sampling plan. It also discusses the key steps in designing a sampling plan, including target population definition, sampling approaches, and determinants of sample size. It also provides basic formulas for calculating sample size for a desired research objective.

Sampling terminology

Several sampling terms are relevant in designing a sampling plan; an understanding of these terms is critical in implementing a sampling plan (Kish 1965; Cochran 1977; Thompson 2002). In sampling, an observation unit is an individual unit from which information is collected. The decision to obtain information from a sample rather than the population is based on several factors, including size of the population, cost of data collection, and error rate. It is possible to observe all units in the population if the population size is small and all observation units are accessible. The cost of data collection also plays a major role. Costs include personnel, time, and resources, and often preclude researchers' collection of data from all observation units in the population. The sampling error rate refers to variability in the measures derived from the sample rather than the population. The error rate decreases as the sample size approaches that of the population. Therefore the measures based on population have zero error rates. Pharmacy practice and policy research is often based on a sample because the data from the sample are generalizable to the population, within the margin of error, at a fraction of the cost.

The goal of research is to obtain a summary measure of interest based on the sample, also referred to as a statistic. Summary measures include mean, median, and mode for the measure of interest, such as prescription volume and price. A summary measure based on all observation units in the population is called a parameter. A well-designed sample will provide a statistic that is consistent with the parameter within the margin of error. A sampling plan is developed before data collection and includes details of target population definition, sampling design, and sample size. Target population includes all accessible units of observation, such as all pharmacies in the state. Sampling design is the process of selection of observation units from the population. The goal of a sampling design is to obtain a representative sample from the target population. A good sampling design provides generalizability of the findings based on a sample. A representative sample means that the sample characteristics reflect population characteristics. Sample size is the number of observation units needed for a sample. Sample size calculations estimate the minimum sample needed to achieve the research objective.

Sampling plan

Sampling plan decisions are based on the research objective and practical considerations. In pharmacy practice and policy research, the research objectives are either descriptive or causal in nature. The research objective determines the target population, sampling design, and sample size. For example, a research study designed to examine dispensing patterns in retail pharmacies within a state should be familiar with the target population, sampling design, and sample size. Practical considerations, such as personnel, costs, and resources, influence the sampling plan. Most research is based on limited resources and budgets. The sampling plan should be developed consistent with the objective within the constraints of personnel, costs, and resources.

Target population

The universe or target population defines the population of interest and identifies observation units in the population. It refers to the theoretical population of interest. It includes all observation units irrespective of accessibility of units for data collection. Accessible population includes accessible observation units from which a sample is derived. Universe and accessible populations can be the same for some research depending on the accessibility. Research objective and practical considerations are the major determinants of the target and accessible populations. The research objective defines the target population, and practical and cost considerations influence the definition of the accessible population. In research involving pharmacies, target and accessible populations can include all registered retail pharmacies in the state. The unit of observation is the retail pharmacy. The sampling unit is the unit selected for sampling that may or may not be the observation unit.

Studies involving multistage sampling comprise multiple stages of selection of sampling units (Lee et al. 1989; Thompson 2002). Sampling units can be geographic areas, states, metropolitan areas, households, or individuals. The first stage of selection in multistage sampling is usually the largest sampling unit, also known as the primary sampling unit, and the last stage is usually the observation unit (Lee et al. 1989; Thompson 2002). The primary sampling units in large national surveys are counties, a few adjacent counties, towns or townships, or metropolitan areas. In the second stage, samples from households can be selected and individuals from within the households can be selected for the final stage. Multistage sampling is often used for operational efficiency. Selection of a representative sample necessitates the listing of all observation units in the population. Multistage sampling provides a process to ensure the representative nature when the listing of observation units is not available. It requires the listing of sampling units for the selection of the primary sampling unit and subsequent sampling units. In multistage sampling, the unit of analysis can be a household or person. The unit of analysis is usually the observation unit.

Sampling design

Sampling design, as mentioned earlier, is the process of selecting sampling units from the population and can be classified as a probability or a nonprobability sample (Kish 1965; Cochran 1977; Thompson 2002). Characteristics of a probability sample include objectivity and random selection. Objectivity precludes biased selection of samples or observation units. Bias in sampling occurs when selection of the sampling unit is based on subjectivity. Probability sampling involves random selection of sampling or observation units and, conversely, nonprobability sampling is characterized by nonrandom selection, convenience, and subjectivity. Nonrandom selection may create an unrepresentative sample and, consequently, findings are less generalizable to the population. Subjectivity, in nonprobability sampling, places the burden of selecting the sample on the researcher. It is, however, convenient and easy to implement. Understanding of the characteristics of the population is vital in selecting an appropriate nonprobability sample.

Research objectives and practical considerations influence sampling plan decisions. Probability sampling allows the researcher to make inferences about the population. The random sampling process is often used for research objectives that require external validity or generalization. These studies extrapolate the findings based on the sample to the population. It is often used in studies involving national surveys; however, probability sampling is cumbersome and costly. There are four types of probability sampling: simple random sampling, systematic sampling, stratified sampling, and cluster sampling (Kish 1965; Cochran 1977; Thompson 2002). Nonprobability sampling is convenient and easy to implement because there are limited restrictions in terms of selection. There are four types of nonprobability samples: convenience sampling, judgment sampling, quota sampling, and snowball sampling (Kish 1965; Cochran 1977; Thompson 2002). Nonprobability sampling is often used with research objectives that are focused on internal validity such as experimental studies.

Probability sampling

Simple random sampling or random sampling is the simplest and most common method of probability sampling. The two key features of simple random sampling are that (1) each unit has a known and equal chance of being selected and (2) each sample has an equal chance of being selected. For example, if there are 1000 pharmacies in a state and 100 pharmacies are selected, each pharmacy has a 1 in 10 chance of being selected, and each sample of 1000 pharmacies has a 1 in 10 chance of being selected. In the simple random sampling method, a complete listing of observation units in the population is needed to select a sample. For example, the 1000 pharmacies in the region may be listed alphabetically. Through the use of a table of random numbers, computer-generated numbers, or by physically selecting 100 pharmacies (e.g., from a hat containing the names of all 1000 pharmacies, after thoroughly mixing them), random sampling can be implemented with replacement and without replacement. In random sampling with replacement, the selected unit is replaced or returned to the population before the subsequent selection. However, the selected unit is not replaced in random sampling without replacement. The sample size and practical considerations dictate whether with or without replacement is needed.

Systematic sampling entails selection of every kth unit from the complete list of observation units in the population to produce a random sample. The selection of the kth unit is based on the number of observation units and sample size needed; k is the quotient or the sampling interval. Using the previous example, if a sample size of 100 is needed from the population of 1000, every tenth pharmacy can be selected; the quotient k is usually obtained by dividing the population by the sample size. The first sampling unit can be selected randomly by choosing a number between 1 and k (10 in this case), e.g., 3. Subsequent selections would then be made of every 10 pharmacies thereafter (13, 23, 33, etc.) until 100 pharmacies have been selected. The type of listing plays an important role in systematic sampling because there is a certain repetitiveness in the selection. If there is any pattern in listing the pharmacies, it would be reflected in the selection. For example, if the listing is by zip (or post) code and not alphabetical, systematic sampling would reflect the distribution of zip codes. If the listing is alphabetical, the sample will be similar to that of simple random sampling. Systematic sampling is simpler and easier to implement than simple random sampling.

Stratified sampling involves random selection of sampling units from each stratum or level of the defined population. Strata are mutually exclusive and exhaustive levels of a categorical variable. The two strata in the metropolitan status of pharmacies are metropolitan and nonmetropolitan. Stratified sampling is designed to provide samples from each stratum level either for comparison or for their representative nature. There are two categories of stratified sampling: proportionate and disproportionate (Cochran 1977; Thompson 2002). Proportionate stratified sampling provides equal probability of selection of sampling units across strata to ensure that the distribution is consistent with the population. If there are more metropolitan than nonmetropolitan pharmacies, this distribution is reflected in the selection of pharmacies from each stratum. A random sample may or may not provide representative sample with respect to strata. Disproportionate stratified sampling implies differing probabilities of selection of sampling units across strata. A higher probability of selection of sampling units exists within the small stratum when compared with strata with higher populations. Sampling weights are needed to account for disproportionate sampling in statistical analysis (Korn and Graubard 1999). Inverse probability weighting is usually used to calculate sampling weights. Disproportionate stratified sampling is advantageous to provide sufficient samples in each stratum for comparative reasons.

Stratified sampling requires listing sampling units within the strata. Using the same example the listing of pharmacies by metropolitan status allows the selection of a stratified sample of pharmacies. The selection of 100 pharmacies from 1000 available pharmacies in the state can be proportional or disproportional. Suppose 80 percent of pharmacies in the population are from a metropolitan location. Proportional sampling will result in selection of 80 pharmacies from 800 pharmacies in the metropolitan stratum and 20 from 200 pharmacies in the nonmetropolitan stratum. Disproportionate stratified sampling will result in a larger sample from the nonmetropolitan stratum when compared with proportional sampling. For example, disproportionate stratified sampling can result in 60 pharmacies from the metropolitan stratum and 40 from the nonmetropolitan stratum. Although this distribution is not consistent with the population, sample sizes are sufficient in both strata for comparative analysis. Sampling weights are used to address the disproportionate sampling in statistical analysis. For metropolitan pharmacies, the probability of selection is 60/800 or 0.075 and the sampling weight will be the inverse of the probability of selection – 1/0.075 or 13.33. Accordingly, the sampling weight for nonmetropolitan pharmacies will be 5.

Cluster sampling involves selection of a group or a cluster of basic sampling units. Clusters contain a group of observation units bound geographically or by other means. Only a random sample of clusters is selected in cluster sampling. All or some of the observation units within the cluster are selected for data collection. Cluster sampling requires the listing of clusters to select a sample. In a single stage sampling, all of the observation units within the cluster are selected after random selection of clusters. In multistage sampling, additional stages of sampling are conducted. As previously mentioned, the largest sampling unit is known as the primary sampling unit, selected in the first stage of selection, whereas the last stage is usually the smallest sampling unit. Multistage sampling can include selection of the primary sampling unit in the first stage and selection of smaller clusters from within the large clusters in the second stage. The final stage can involve selection of observation units within the small clusters. Cluster sampling is advantageous when the listing of observation units is not available or when observation units are geographically dispersed.

In cluster sampling, geographically bound clusters, such as geographic locations or zip codes, reduce the cost of data collection. For example, each zip code can be considered a cluster. Proportional sampling is usually beneficial to prove a representative sample of pharmacies, creating a proportion consistent with the distribution in the population. For example, if there are 20 zip codes in a state, 10 zip codes can be randomly selected. In single-stage sampling, all the pharmacies within a given zip code area are selected. In multistage sampling, the first stage can involve selection of zip codes and the second stage selection of a sample of pharmacies within a zip code area. Cluster sampling is beneficial if a pharmacist in each pharmacy is needed for an interview. Cluster sampling will help to reduce the travel time and cost of data collection by selecting pharmacies within the zip code area. Cluster sampling is also useful if a listing of pharmacies is readily available but not the listing of pharmacists.

Nonprobability sampling

Convenience sampling is a nonprobability sampling process wherein the observation units are selected based on the availability of participants and the researcher's convenience. It is easy to implement because a sample is selected in locations, such as universities, shopping malls, or healthcare institutions, where there is a possibility of identifying a large group of individuals and it is convenient for the researcher. Convenience samples may not represent the population and consequently the findings are only generalizable to the selected sample. Convenience sampling is often used in exploratory studies, pilot studies, or studies that require pre-testing. It is also used in randomized clinical trials (RCTs) because the goal of such studies is to compare safety and efficacy outcomes in the treatment and placebo group rather than to obtain a sample that is generalizable to the population. It is easy to implement but serves limited research purposes. A sample of pharmacies that is available and convenient to a researcher can be selected to address dispensing issues. Findings from such a study are generalizable only to the selected pharmacies.

Quota sampling is a form of stratified sampling without the need for random selection of sampling units from the target population. It involves selection of a convenient study sample with the desired characteristics. The desired characteristics are often sociodemographic variables, such as gender, ethnicity, or location. The process requires identification levels of a categorical variable to provide the desired sample. The quota or distribution of desired characteristics for the convenient sample is often based on the population distribution. Consequently, quota sampling requires data regarding the characteristics of the population. Supposing that there are 80 percent of pharmacies in the population from a metropolitan location, quota sampling based on metropolitan location will result in a convenient sample of 80 metropolitan pharmacies and 20 nonmetropolitan pharmacies. The biggest difference between stratified and quota samples is nonrandom selection. Although a quota sample may represent some characteristics of the population, the findings are not generalizable. Quota sampling is often used to select a sample with the desired characteristics for convenience and cost reasons.

Judgment sampling is a nonprobability sampling process designed to identify a sample that, in the researcher's or expert's judgment, will serve the research purpose. It does not involve identification of desired characteristics before the selection. It is the most subjective sampling process because the decision to identify is based on an opinion. Although it is subjective, experienced researchers and experts can have a strong understanding of representative samples. The judgment sample is often selected for exploratory studies and it is often used in pilot studies or studies that require pre-testing. Judgment sampling tends to provide better samples than convenient sampling because the selection criteria are based on an expert opinion. For example, an interview instrument can be tested using a judgment sample of pharmacists who, in the researcher's opinion, represent all pharmacists in the state. The findings based on judgment samples are not generalizable even though they are considered representative by the researcher.

Snowball sampling starts with a very small convenient sample that in turn helps to recruit more respondents. The sampling process is analogous to a rolling snowball as the initial participants help to attract relevant participants for the research. Snowball sampling is often used to recruit participants who are difficult to identify in large numbers, such as pharmacists involved in complementary medicine. All participants in snowball sampling act like recruiters. Snowball sampling tends to provide samples that are similar in some basic characteristics of interest. It is a subjective process and the sample does not represent the population. However, it is an effective way to identify a large number of participants who are difficult to locate or recruit for a study.

Randomization and random sampling

Both random assignment and random sampling are based on the concept of probability theory. However, the purpose and the goals of these are different. Experimental studies involve randomization or random assignment of participants to treatment and control groups to address cause-and-effect relationships between an intervention and an outcome. Random assignment involves random allocation of selected participants or patients to experimental or control groups. This is based on the concept that selected participants or patients have an equal chance of being selected into the experimental or control group. The goal is to minimize bias in assigning the intervention and to make the groups equivalent or comparable with respect to measured and unmeasured characteristics that can influence the outcomes, as discussed in Chapter 5. Random assignment distributes participants or patients with varying characteristics approximately equally, which ensures that the effect of measured and unmeasured characteristics on the outcome is equal in the randomly assigned groups and any differences in the outcome can be truly attributed to the intervention (Kerlinger 1986).

On the other hand, random sampling is based on the concept of external validity. It addresses the question: Are the findings generalizable? Random sampling involves selection of a representative sample from the population. This ensures that the selected sample represents the population and the findings based on the sample can be extrapolated to the population. A truly scientific study designed to address causal relationships should involve random sampling and random assignment to address both internal and external validity. This can be achieved by selecting a random sample from the population, and then selected patients or participants are given random assignments to an experimental or control group, often referred to as random selection, followed by random assignment to an intervention (Mikeal 1980). The other option is to select two equal and independent random samples from the population and randomly assign them to an intervention. This is sometimes referred to as the random sampling model (Hsu 1989). Both approaches provide similar results with respect to equal distribution of characteristics. Practically speaking, most clinical trials involve nonprobability sampling followed by random assignment to treatment because the focus is more on internal validity than on external validity. Often the nonprobability sampling in RCTs involves convenient samples or some kind of quota sampling to enhance the generalizability.

Sample size

Sample size calculations are important in research for various scientific and practical reasons. Estimated sample size ensures that findings from the sample are generalizable within the margin of error. Sample size information is vital for estimating the resources, time, and budget needed to conduct the research. Consequently, sample size calculations should be conducted as part of the sampling plan. Several factors influence the sample size calculations, including study objective, study design, practical considerations, and statistical analysis to be performed. Sample size calculations are estimations based on a given set of factors and can change if there is any change in the determinants. Estimated sample size based on these factors provides scientific and practical value in conducting research.

The study objective defines the formula for sample size calculations. The determinants in the formula influence the sample size. Basic formulas for sample size calculations estimate the population parameter and differences within two groups (Kish 1965; Cochran 1977; Thompson 2002). Studies designed to estimate population parameters are often focused on external validity. The following are examples of research questions dealing with population parameters:

- What is the mean prescription volume for behind-the-counter products per day for pharmacies in the state?
- What percentage of pharmacies has a drive-through prescription pick-up provision?

These research questions estimate population mean or proportion based on a sample. The sample size calculations provide the minimum sample size needed to implement probability sampling design.

Studies designed to compare group differences are focused on internal validity. Research questions for estimating differences in two groups include:

- Is there a difference in the mean prescription volume per day between metropolitan and nonmetropolitan pharmacies?

- Is there a difference in proportion of pharmacies in metropolitan and nonmetropolitan pharmacies offering a drive-through prescription pick-up option?

These research questions compare means or proportions in two samples. The sample size calculations provide the minimum sample size needed to implement random sampling and/or random assignment.

Sample size for estimating population mean

Research questions estimating population mean based on a random sample require an approximate standard deviation and margin of error to calculate the sample (Ott 1992):

$$n = \frac{(Z_\alpha \times \sigma)^2}{e^2}$$

where n = sample size needed, Z_α = Z score for α or type I error, σ = standard deviation, and e = margin of error.

The Z_α score is generally used for calculating the confidence interval of a mean. It is based on the central limit theorem. Most calculations are based on α or type I error of 5 percent to provide a 95 percent confidence interval. The other confidence coefficients are 90 and 99 percent. Type I error or α is an error in rejecting a null hypothesis when it is true. Intervals calculated using the Z_α score (two-sided) of 1.96 will contain the population mean 95 percent of the time. The Z_α score (two-sided) for a 99 percent confidence interval is 2.57. A decrease in α or an increase in Z score increases the sample size.

Standard deviation refers to the variability of a measure in the population. Standard deviation is calculated based on the differences between individual measures and the mean. A high standard deviation indicates that the individual measures are widespread or highly dispersed in the population. A low standard deviation indicates that the individual measures are close to each other. The higher the standard deviation or variability in the measure, the larger the sample size needed. Standard deviation for sample size calculation can be approximated based on past studies, pilot study, or judgment, or by dividing the range by 4 (Ott 1992).

Margin of error, also referred to as sampling error, is expressed as the actual value of the error that is acceptable in calculating the population mean. The margin of error varies with the research objective. As the margin of error score decreases, the sample size increases. For example, the following are needed to estimate mean prescription volume for behind-the-counter products per day for pharmacies in the state: an approximate standard deviation (5), an acceptable margin of error (±2 prescriptions), and a Z_α (1.96) score. Using the above values, the sample size would be 24.01 or 25 pharmacies.

An appropriately calculated random sample represents the population irrespective of the size of the population. Generally, the size of the population does not play a role in sample size calculations. A correction factor is needed only if the sample constitutes a large portion of the population. This is known as the finite population correction (FPC) factor (Cochran 1977; Thompson 2002). It measures the increase in precision or the decrease in error when sample size approaches the population size. The quotient of the FPC calculation provides the corrected sample size:

$$FPC = \sqrt{\frac{N-n}{N-1}}$$

where FPC = finite population correction factor, N = size of the population, and n = size of the sample.

Sample size for estimating population proportion

Research questions estimating population proportion based on a random sample require a proportion estimate and an acceptable percentage of margin of error in order to calculate the sample size (Ott 1992):

$$n = \frac{Z_{\alpha}^2}{e^2} \times (\pi) \times (1 - \pi)$$

where n = sample size needed, Z_{α} = Z score for α or type I error, π = proportion estimate, and e = percentage of margin of error.

Although the research intends to estimate population proportion, an approximate proportion estimate is needed for sample size calculation. It can be approximated based on past studies, pilot study, and researcher's judgment. If it cannot be approximated, use of 0.5 as a proportion estimate will provide the largest possible sample size. A highly precise study may require a ±1 percent level of error, whereas other studies can accept a high margin of error such as ±5 percent. This percentage is often reported in public polls. For example, the following are needed to estimate the percentage of pharmacies with drive-through prescription pick-up provision: an approximate proportion estimate (50%), an acceptable percentage margin of error (±5%), and a Z_{α} (1.96) score. Using these values, the sample size would be 384.16 or 385 pharmacies.

Sample size for comparison of population mean

Research questions comparing two population means based on a sample, require an approximate mean of the two groups or the difference in the mean of the two groups and the standard deviation (Ott 1992):

$$N = 2\frac{(Z_{\alpha}+Z_{\beta})^2 \times \sigma^2}{(\mu_1 - \mu_2)^2}$$

where $N=$ sample size needed, $Z_\alpha = Z$ score for α or type I error, μ_1 and $\mu_2 =$ means in the two groups, $Z_\beta = Z$ score for β or type II error, and $\sigma =$ common standard deviation.

The Z_α score interpretation is the same as in previous formulas. The Z_β score represents the power component in the sample size calculation. The power of a study is the probability that the study would reject a null hypothesis when it is false. It refers to the probability of detecting a difference between the groups in the sample, if there is a difference in the population. Type II error or β is an error in acceptance of a null hypothesis as true when it actually is false. Power is $1 - \beta$ and is 0.84 for a power of 80 percent and β of 20 percent. The Z_β score (one-sided) for 90 percent power is 1.28. As the power or Z_β score increases, the sample size also increases, and an increase in power means that the probability of detecting a difference is high if there is a difference between the two groups. For example, the following are needed to compare two means involving patients with high cholesterol: the approximate total cholesterol or difference in two groups (200 and 190 mg/dL), standard deviation (10 mg/dL), Z_α (1.96) score, and Z_β (1.28) score. Using these values, the sample size would be 21 per group.

Sample size calculations ensure that findings derived from appropriate samples are clinically and statistically significant. Statistically significant findings with limited clinical significance have little or no value in pharmaceutical practice and policy. Consequently, the difference in the mean in the two groups specified for sample size calculations should be clinically meaningful for pharmaceutical practice. The smaller the difference between the groups, the bigger the sample size needed to detect the difference. Similarly, a smaller size is needed to detect a big difference in the two groups. Irrespective of the size of the difference, the value has to be clinically significant to be meaningful in practice. The effect size is the difference in the mean divided by the standard deviation. It refers to the standardized difference in the mean of the two groups, also known as Cohen's d (Cohen 1988). The standardized mean difference adjusts for the variability in the measures across the sample.

Sample size for comparison of population proportion

Research questions comparing two population proportions based on a sample require approximate proportions in the two groups (Ott 1992):

$$N = \frac{\left[Z_\alpha \times \sqrt{2\pi_1(1-\pi_1)} + Z_\beta \times \sqrt{\pi_2 \times (1-\pi_2) + \pi_1 \times (1-\pi_1)} \right]^2}{(\pi_1 - \pi_2)^2}$$

where N = sample size needed, $Z_\alpha = Z$ score for α or type I error, π_1 and π_2 = proportions in the two groups, and $Z_\beta = Z$ score for power or $1 -$ type II error

The interpretations of Z_α and Z_β are the same as in previous formulas. The proportions in the two samples are needed for sample size calculations. It can be based on past studies, pilot studies, and researcher's judgment. Similar to the previous equation, a larger sample size detects small difference in proportions between the two groups. The difference specified should be clinically meaningful for research to have implications in practice. For example, the following are needed to compare proportions of patients that reach desirable cholesterol levels in two groups: the approximate proportions or difference in the two groups (0.90 and 0.50), Z_α (1.96) score, and Z_β (1.28) score. Using these values, the sample size would be 52 per group.

Other considerations

The formulas discussed above are primarily intended to provide a basic understanding of determinants of sample size. It is not a comprehensive list of all formulas for sample size calculations for various research questions. Different sample size formulas are available to address various types of study objectives. The statistical determinants generally include α and β and these are often standardized. The other determinants, such as standard deviations, means, and proportions, are user specified, and decisions about these determinants should be based on the research objective and previous research. In general, conservative estimates should be used to provide maximum sample sizes. Sample size calculations are estimations and may be altered due to changes in underlying assumptions. Most basic sample size calculations are based on the assumption of normal distribution. Samples derived from other distributions of values require complex formulas. Most software packages such as SAS and STATA have procedures to calculate sample sizes (SAS Institute 2006; StataCorp 2007). These procedures calculate sample size and power based on the type of research question and statistical analysis to be performed. The data needed to calculate sample size in the statistical packages are similar to the basic formulas discussed above. Other determinants of sample size calculations vary with the procedures. Previous research and pilot findings can be useful in estimating sample size based on any formula.

Study design considerations

Research objectives determine the structural framework for study implementations also known as the study design. It includes details of stimulus and response, time ordering of stimulus and response, and sampling and allocation procedures. Experimental research involves a manipulated stimulus whereas the stimulus is observed only in nonexperimental studies. Experimental studies usually require a smaller sample size than nonexperimental studies. The temporal ordering of stimulus and response in research can be varied based on the needs of the experimental and nonexperimental research. In clinical research, designs such as parallel and crossover designs are popular. Crossover designs typically require a smaller sample than parallel designs, due to repeated use of the same sample. In nonexperimental research, case–control designs usually require a smaller sample size than cohort studies. Most sampling calculations are based on the assumption of simple random sampling. Studies that require stratified and cluster sampling influence the sample size calculations. Sample size procedures available in most statistical packages can be used to obtain samples for different study designs.

Practical considerations

The data collection issues, such as response rate and resources, influence the sample size needed for research. Response rate refers to the percentage of participation from the available respondents. Several factors influence the response rate such as research topic, data collection methodology, and characteristics of the respondents. Respondents' interest in the research topic generally lead to high response rates. Similarly, data collection processes that require less time and are less intrusive have high response rates. Longitudinal studies usually require large sample sizes due to attrition or dropout rates. Studies involving multiple survey items may have missing or incomplete responses and, thereby, influence the effective response rates of such studies. Respondents' characteristics such as gender, age, and education also play a role in influencing the response rates. These factors have to be considered in determining the effective sample size, appropriately adjusted for the response rate. The response rates are approximations and should be conservatively estimated to obtain the maximum sample. Previous studies, pilot studies, and experience can be helpful in estimating the response rate before the study. For example, if the response rate is 50 percent from the pharmacies, twice the calculated sample size is needed to obtain the desired sample size:

$$n_1 = \frac{n_2}{(1-q)}$$

where n_1 = effective sample size, n_2 = calculated sample size, and q = response rate.

Other practical consideration, such as budgets, personnel, and time for data collection, can influence the effective sample size. Reduction in effective sample size reduces the power and, consequently, decreases the probability of detecting a difference if there is in fact a difference in the two groups. Response rates have to be managed to maximize the sample size.

Analytical considerations

The sample size selected should be consistent with the statistical analysis to be performed on the data. Although study objective and analysis are interlinked, statistical tests have their own requirements. These considerations should be reflected in the sample size. For example, Z tests require a minimum sample size of 30 due to sampling distribution. A sample size of 30 will not ensure that the sample is right for a research objective. Rather, it will ensure only that the analysis is right for the sample size. Consequently, statistical tests are just a part of the sample size considerations. A sample size of 30 is often considered the minimum cell size in studies involving multiple factors when sample size calculations are not possible. The number of variables and levels of variables to be used in the analysis influence the statistical analysis, and consequently the sample size, because most sample size formulas are derived from the formulas of statistical tests. Most sample size considerations are driven by the primary stimulus and response variables of interest.

The type of response variable influences the statistical analysis and the formula to be used for sample size as discussed previously. Similarly, the number of levels in the stimulus also influences statistical tests used and sample size calculations. Studies that require subgroup analysis should ensure that sample sizes are sufficient to conduct such analysis. If the researcher is interested in comparing chain and independent pharmacies in metropolitan areas, the sample sizes in each of the subgroups should be sufficient to conduct such AN analysis. Studies that require regression analysis (multiple, logistic, or survival analysis) to control for multiple factors can be based on stimulus and response variables of interest. The sample size calculations for such analyses can be based on basic sample size calculations adjusted for multiple factors by using a variance inflation factor (Hsieh et al. 2003). The general rule of thumb can also be used; for example, a minimum of 10 observations is needed per predictor variable in regression analysis (Hosmer and Lemeshow 2000). There are rules of thumb for other statistical analyses (VanVoorhis and Morgan 2007).

Summary and conclusions

Sampling plans are important in pharmacy practice and policy research for practical and scientific considerations. Well-designed research should include details of target population, sampling approaches, and sample size. The research objective is the primary driving force in sampling plan decisions.

Practical considerations such as personnel, costs, and time also influence the sampling plan. The key factor in sampling is defining the target population. The sampling approaches can be grouped into probability and nonprobability samples. Probability sampling includes simple random sampling, systematic sampling, stratified sampling, and cluster sampling. Nonprobability sampling includes convenience sampling, judgment sampling, quota sampling, and snowball sampling. The criteria for selecting sampling approaches are based on internal and external validity issues. The sample size needed for a study is based on study objective, study design, practical considerations, and the statistical analysis to be performed. These sample size considerations are designed to provide a basic understanding of determinants of sample size calculations. Sample size calculations can be altered by changing the determinants. Sampling plan development should be in accordance with the research objective within the constraints of personnel, costs, and resources. An appropriate sample will ensure the scientific and statistical validity of the research.

Review questions/topics

1 Discuss probability sampling approaches using an example.
2 Discuss nonprobability sampling approaches using an example.
3 What are the basic formula determinants in sample size calculations?
4 Discuss how study design and statistical considerations influence sample size.
5 Discuss how practical considerations influence sampling plans.

References

Cochran WG (1977). *Sampling Techniques*. New York: Wiley.
Cohen J (1988). *Statistical Power Analysis for the Behavioral Sciences,* 2nd edn. Hillsdale, NJ: Lawrence Earlbaum Associates.
Hosmer D, Lemeshow S (2000). *Applied Logistic Regression,* 2nd edn. New York: Wiley.
Hsieh FY, Lavori PW, Cohen HJ, Feussner JR (2003). An overview of variance inflation factors for sample-size calculation. *Eval Health Prof* 26: 239–57.
Hsu LM (1989). Random sampling, randomization, and equivalence of contrasted groups in psychotherapy outcome research. *J Consult Clin Psychol* 57: 131–7.
Kerlinger FN (1986). *Foundations of Behavioral Research*, 3rd edn. New York: Holt, Rinehart & Winston.
Kish L (1965). *Survey Sampling*. New York: Wiley.
Korn EL, Graubard BI (1999). *Analysis of Health Surveys*. New York: Wiley.
Lee ES, Forthofer RN, Lorimer RJ (1989). *Analyzing Complex Survey Data*. Newbury Park, CA: Sage Publications.
Ott RL (1992). *An Introduction to Statistical Methods and Data Analysis*. Belmont, CA: Wadsworth.
Mikeal RL (1980). Research design: general designs. *Am J Hosp Pharm* 37: 541–8.
SAS Institute (2006). *SAS 9.1.3 Procedures Guide*, Vol 4. Cary, NC: SAS Institute.
StataCorp (2007). *Stata Statistical Software: Release 10*. College Station, TX: StataCorp LP.

Thompson SK (2002). *Sampling*. New York: Wiley.

VanVoorhis CR, Morgan BL (2007). Understanding power and rules of thumb for determining sample sizes. *Tutorials Quant Methods Psychol* 3: 43–50.

Online resources

Centers for Disease Control and Prevention. "Epi Info." Available at: www.cdc.gov/EpiInfo.

Department of Biostatistics. Vanderbilt University. "PS: Power and Sample Size Calculation." Available at: http://biostat.mc.vanderbilt.edu/wiki/Main/PowerSampleSize.

Division of Mathematical Sciences, the College of Liberal Arts or the University of Iowa. "Java applets for power and sample size." Available at: www.stat.uiowa.edu/~rlenth/Power.

World Health Organization. "STEPS Sample Size Calculator and Sampling Spreadsheet." Available at: www.who.int/chp/steps/resources/sampling/en/index.html.

8

Systematic review of the literature

Darren Ashcroft

Chapter objectives

- To present types of literature reviews
- To describe the steps involved in undertaking a systematic review
- To implement search strategies to locate research evidence
- To abstract study findings qualitatively and quantitatively
- To assess the quality of systematic reviews

Introduction

Evidence of the effectiveness of healthcare interventions should ideally be used to inform decisions about the organization and delivery of healthcare services relating to pharmaceutical practice and policy. However, the medical and pharmacy literature is expanding at an increasingly rapid rate. Reviews of research evidence can help increase access to the existing knowledge base by ordering and evaluating the available evidence. Reviews are important to understand the current state of knowledge and are instrumental in planning future research.

This chapter focuses on systematic reviews to provide the reader with a practical guide to understand and appraise this type of study. Initially, a brief overview of review terminology is provided before outlining the benefits of systematic review methods. The chapter then moves on to consider how to conduct and then subsequently locate and evaluate the quality and results of systematic reviews.

Review terminology

Literature reviews can vary considerably in design, execution, and assessment of the evidence and, in practice, the terminology used to describe different types of reviews has also varied. Although the term "meta-analysis" is often used interchangeably with systematic review, strictly speaking it is a statistical technique used to combine results from a series of different studies into a single summary estimate. Over many years of application and development, meta-analysis has been established as a valuable method for secondary research with its own systematic procedures and methodology for evaluating study design, sampling, and analysis. A meta-analysis is also often referred to as a quantitative systematic review. In contrast, when the results of primary studies are summarized but not statistically combined, the review is often called a qualitative systematic review. Summaries of research that lack explicit descriptions of systematic methods are often called traditional narrative reviews.

Traditional, or narrative, reviews are usually compiled by experts in the field, and may range in scope from comprehensive overviews to discussions of a focused question. Not surprisingly, they may be biased by the author's awareness or preference for particular studies and the methods of synthesizing information from different studies. Systematic reviews differ from other types of literature review in that they adhere to a strict scientific design in order to make them more comprehensive, to minimize the chance of bias, and so ensure their reliability. Rather than reflecting the views of the authors or being based on only a (possibly biased) selection of the published literature, they contain a comprehensive summary of the available evidence.

Importance of systematic reviews

There are two main practical reasons for the importance of systematic reviews. First, the systematic approach used to identify and select all relevant evidence minimizes the threat of bias in the interpretation of the existing evidence base. Some of the most common types of selection bias associated with literature reviews are outlined in Table 8.1. As most reviews depend on

Table 8.1 Common types of selection bias in literature reviews

Publication bias	Selective publication of studies with positive findings or statistical significance
English language bias	Selective submission of positive or statistically significant studies to major English language journals
Database bias	Biased indexing of published studies in literature databases
Outcome bias	Several outcomes within a study are measured but these are reported selectively depending on the strength and direction of the results
'Gray' literature bias	Results reported in journal articles are systematically different from those presented in reports, working papers, dissertations, or conference abstracts

published studies, the most important is often publication bias, which refers to the selective publication of studies that show positive results or statistical significance, although other forms of bias may be equally important for particular types of reviews.

Dwan et al. (2008) reviewed and summarized the empirical evidence of publication bias and outcome reporting bias in randomized controlled trials (RCTs). Based on a series of cohort studies reporting on bias in RCTs, they found strong evidence of an association between significant results and publication. Studies that reported positive or statistically significant results were more likely to be published and, when comparing trial publications with protocols, 40–62 percent of studies had at least one primary outcome that had been changed, introduced, or omitted. Despite recent requirements for investigators to register clinical trials as a precondition to publishing a trial's findings, Mathieu et al. (2009) have also shown that selective outcome reporting remains a major problem, which may adversely affect the validity of any systematic review based on incomplete evidence. It is important that authors of systematic reviews are aware of these potential sources of bias, minimize the impact of them whenever possible, and investigate and report on their likely presence within the final review report.

The second main reason for the importance of systematic reviews relates to the fact that smaller studies may have a rigorous study design, but lack the statistical power to detect a statistically significant effect. The added power brought about by synthesizing the results of a number of smaller studies may lead to the identification of earlier conclusions to help inform decisions about the organization and delivery of healthcare.

In summary, systematic reviews locate, appraise, and synthesize evidence from research studies in order to provide informative empirical answers to scientific research questions. They are, therefore, valuable sources of information for decision makers, such as healthcare practitioners, researchers, and healthcare funding organizations. In addition, by identifying what is and what is not already known, they are an invaluable first step before carrying out new primary research.

Getting started

Is a review required?

Traditionally, in healthcare, systematic reviews have been undertaken to establish the clinical effectiveness and/or cost-effectiveness of an intervention or drug therapy. Organizations such as the Cochrane (www.cochrane.org) and Campbell Collaborations (www.campbellcollaboration.org) provide considerable support to researchers interested in conducting systematic reviews, especially in combining numerical data from experimental studies in the form of a meta-analysis to answer questions about "what works?" The Agency for Healthcare Research and Quality (AHRQ) has also developed a methods guide for the conduct of comparative effectiveness reviews (see Online resources at end of chapter).

Systematic reviews may also be needed to propose a future research agenda when future research direction is unclear or existing agendas have failed to address a specific problem. As a result, they now often feature in student dissertations or postgraduate theses, and may be required by research teams who wish to secure grant funding for new primary research.

Increasingly, systematic reviews are also being undertaken to determine whether an intervention is feasible or appropriate, or relates to evidence of experiences, values, and beliefs of patients and/or healthcare practitioners. So, although all systematic reviews use formal, explicit methods to describe and synthesize evidence, they can vary considerably in the types of questions that they aim to answer. Different types of evidence will be suitable for answering different questions, and different methods will be appropriate for synthesizing different types of evidence.

Development of a review protocol

The review protocol outlines the predetermined plan that the systematic review will follow. A review is less likely to be biased if the research question is clearly defined and the methods to be used to answer it are determined before gathering the relevant data and drawing conclusions. If not, it is possible that study selection may be unduly influenced by assumptions about the likely findings.

The protocol should state in detail the main question(s) that will be examined in the review, along with details of the proposed search strategy, approaches that will be used to critique the robustness of the identified studies, and an outline of the methods that will be used to extract and synthesize the relevant evidence. At the outset, it may not be possible to state explicitly which methods will be used to synthesize the data until after the studies have been assessed, but the approaches that are likely to be used should be specified in the protocol. In addition, any checklists that will be used to assess the validity of the studies should also be included, along with the number of individuals who will independently assess validity, and details of the process by which any disagreements in data extraction will be resolved.

There is no single protocol format, nor is there an accepted set of contents that always needs to be followed. Organizations, however, such as the Cochrane Collaboration have produced a protocol template (as outlined in the *Cochrane Handbook*) that may be suitable to adapt to meet the specific needs of a new systematic review. Key elements that should be addressed in systematic review protocols are summarized in Box 8.1.

Box 8.1 Key sections of a systematic review protocol

Title of the systematic review
Protocol information:

Authors
Date of protocol (version number)

The protocol:

Background to the review question
Objectives of the review

Methods
Criteria for selecting studies for this review:

Types of studies
Types of participants
Types of interventions
Types of outcome measures

Search methods for identification of studies
Details of proposed data extraction and analysis

Undertaking the review

Undertaking a systematic review typically follows five fundamental steps:

1 formulating the research question
2 conducting the literature search
3 refining the search by applying predetermined inclusion and exclusion criteria
4 extracting the relevant data and assessing study quality
5 synthesizing, interpreting, and reporting the data.

An example of a study that follows these five steps is described in Table 8.2.

Formulating the research question

Formulating the right question requires a clear statement of the objectives of the review. Is it a question about the efficacy of an intervention, incidence of an illness, or prognosis, or an economic question? Different types of health-care questions often require different study designs, and consequently different methods of systematic review. Table 8.3 provides an overview of the most appropriate study designs for answering particular types of clinical or public health questions.

Table 8.2 Practical guide to conducting a systematic review[a]

Formulating the question

The population of interest was identified as adult and/or children hospital inpatients

Outcomes of interest included prevalence and incidence rates of prescribing errors on handwritten prescriptions and type of prescribing error

The study designs chosen to be included in the systematic review were randomized controlled trials, nonrandomized comparative studies, and observational studies

Conducting the literature search

Computerized searches of MEDLINE, EMBASE, CINAHL, and International Pharmaceutical Abstracts were used to identify eligible studies published between 1985 and 2007. Several key words were used in the search. Bibliographies of all included studies were also examined to capture studies not identified from the electronic searches

Refining the literature search

Only original research studies were evaluated. In addition, studies that evaluated errors for only one disease or drug class or for one route of administration or one type of prescribing error were excluded. The paper reported on how many studies were eliminated at each step and the reasons for elimination. Overall, 63 publications were included in the systematic review, reporting on 65 unique studies

Data extraction

Data to be extracted were formulated during the design stage of the study: year and country, study period, hospital setting, methods (including study design, sampling and error validation processes, profession of data collector, means of detecting error), error definitions used, frequency of different types of errors, and the error rate reported (or calculated from available data). Data were extracted independently by two of the investigators. The quality of the studies was not rated using a numerical scoring scale due to the wide range of different study designs included

Data synthesis

The summary outcomes included median (and interquartile range) incidence and prevalence rates (error rate of all medication orders, errors per 100 admissions, and errors per 1000 patient-days), types of prescribing errors detected and the medications involved. Because of variations in the definition of a prescribing error, the methods used to collect error data and the setting of the studies, meta-analysis was not undertaken

[a] This example is taken from a systematic review of prescribing errors in hospital inpatients.
Source: Lewis et al. (2009).

In defining an appropriate review question, it can be helpful to answer the following four questions:

1 What is the population of interest?
2 What are the interventions being considered?
3 What are the outcomes of interest?
4 What study designs are appropriate to answer the question?

In practice, not all the questions may be of relevance to all systematic reviews, and many reviews will extract data from a range of different study

Table 8.3 Ideal study designs for research questions

Question type	Ideal study design
1. Intervention	Randomized controlled trial
2. Frequency/rate (e.g., burden of illness)	Cross-sectional study
3. Etiology and risk	Cohort study
4. Prediction and prognosis	Cohort study
5. Diagnostic accuracy	Random or consecutive sample
6. Understanding human behavior	Qualitative research
7. Cost-effectiveness	Economic evaluation

designs. For example, in the systematic review outlined in Table 8.2, the question of interest related to the burden of prescribing errors in hospital inpatients and as a result no specific interventions were being considered, but data to inform the review were drawn from a variety of different study designs. For clinical questions, it is necessary to define the population (participants), types of intervention (and comparisons), and types of outcomes that are of interest. The acronym PICO (participants, interventions, comparisons, and outcomes) is often used as an aide memoire in formulating such questions. Issues to consider relating to the intervention and relevant comparators include possible variations in dosage/intensity, mode of delivery, personnel involved, frequency of delivery, duration of delivery, or timing of delivery. If such variations exist, it is important to consider whether all will be included in the review or whether there is a critical dose below which the intervention may not be clinically appropriate (Higgins and Green 2008).

Conducting the literature search

The objective of a systematic review is to evaluate studies that address the question of interest comprehensively and systematically. To achieve this, it is important that all such studies are retrieved. Initially, it is helpful to find out if a systematic review is under way or has been completed. If not, original studies need to be found that specifically address the question of interest.

Typically, the selection of information sources is often guided by issues such as the discipline(s) of a particular study, the types of studies to be included (quantitative and/or qualitative), and the period of time to be covered. It may be necessary to systematically search:

- general medical, pharmacy, or nursing databases (e.g. MEDLINE, EMBASE, International Pharmaceutical Abstracts (IPA), and/or CINAHL)
- specialist subject databases (e.g. Cancerlit)
- databases from other disciplines (e.g. Applied Social Sciences Index and Abstracts [ASSIA], PsycInfo, and/or the Social Science Citation Index)
- databases of existing reviews (e.g. the Cochrane Library).

Box 8.2 provides further details concerning the main general medical and pharmacy electronic databases. The degree of overlap of literature among these databases varies widely according to the topic, but studies comparing searches of the MEDLINE and EMBASE databases have generally concluded that a comprehensive search requires that both databases be searched (Suarez-Almazor et al. 2000). Examining the search strategies of earlier systematic reviews on related topics may help in constructing an efficient review search strategy. For example, when addressing clinical questions, the search strategy will typically have three sets of terms included: (1) terms to search for the health condition of interest, i.e., the population; (2) terms to search for the

---

Box 8.2 Characteristics of the main general medical and pharmacy electronic databases

- MEDLINE currently contains over 16 million references to journal articles from the 1950s onwards. Currently, 5200 journals in 37 languages are indexed for MEDLINE: www.nlm.nih.gov/pubs/factsheets/medline.html
- PubMed provides access to a free version of MEDLINE that also includes up-to-date citations not yet indexed for MEDLINE: www.nlm.nih.gov/pubs/factsheets/pubmed.html
- EMBASE currently contains over 19 million records from 1974 onwards. Currently 7000+ journals are indexed for EMBASE in 30 languages: www.info.embase.com
- International Pharmaceutical Abstracts (IPA) is produced in cooperation with the American Society of Health-System Pharmacists and currently contains over 350 000 records from 1970 onwards. Each year approximately 10 000 new records are added covering many pharmacy journals and abstracts from major pharmacy meetings: www.ovid.com/site/catalog/DataBase/109.jsp?top=2&mid=3&bottom=7&subsection=10

intervention(s) evaluated; and (3) terms to search for the types of study design to be included (such as using a study "filter" to select only RCTs). Organizations such as the Cochrane Collaboration have developed highly sensitive search strategies (study "filters") for the identification of RCTs in electronic databases such as EMBASE and MEDLINE, which are available online in the *Cochrane Handbook* (Higgins and Green 2008).

It is usually easy to find a few relevant studies by a straightforward literature search. In contrast, finding all relevant studies that have addressed a single question is not easy. At the outset, it is important to define the disciplines and types of studies to include, and it may be helpful to work with an information specialist in identifying the relevant sources for these. In practice, a number of databases are often searched using a customized search filter, which is often supplemented by reviewing the reference lists of all eligible studies to identify any additional work that may not have been indexed by the electronic databases and/or contacting experts in the field of interest.

Refining the search

Not all the studies retrieved from a literature search will answer the question of interest. It is, therefore, necessary to establish inclusion and exclusion criteria before data extraction in order to avoid the risk of selection bias. These criteria should flow logically from the research question under investigation. Often, it is helpful to pilot the criteria to ensure that they are reliably interpreted and that they classify the studies appropriately. Selection criteria typically define:

- the health intervention of interest and any relevant comparisons
- the setting and relevant populations (such as patients, practitioners, or client groups)
- the eligible study designs
- the outcomes of interest.

The initial search for studies is often broad and inclusive and, once the list of studies has been compiled, the titles can be compared against the inclusion and exclusion criteria. Typically, the abstracts of studies that pass the title search are examined and nonrelevant articles can then be excluded. The complete text of the remaining studies is then examined, and those not meeting the inclusion criteria are eliminated. It is important to keep track of the studies retrieved at each stage and the number excluded under each inclusion or exclusion criterion. Often this process is summarized in a "study flow diagram" presented in the published review.

For the systematic review on prescribing errors (outlined in Table 8.2), the electronic searches identified 595 publications, of which 493 were excluded on the basis of their title and published abstract. The remaining 102 publications were obtained in full text and assessed for suitability, as shown in the study flow diagram (Figure 8.1). Further evaluation of the selected publication led to deletion of 51 publications; these studies did not meet the predefined inclusion criteria or they were duplicate publications of the same study. Screening the reference lists of the included publications identified 12 additional eligible studies. Overall, 63 publications were included in the systematic review, reporting on 65 unique studies.

Data extraction

As in any other data-gathering exercise that requires interpretation, observers may disagree. For example, when studies list a variety of patient subgroups, outcome measures, or exclusions, it is possible that readers may vary in how they interpret the data from a particular study. With this in mind, the data to be extracted should be agreed upon by consensus at the start during the design stage of the review. Systematic abstracting of the study characteristics, sample demographics, and reported findings using a structured data extraction form is necessary to enable cross-checking against data extracted by other reviewers within the research team. This can also ensure that any missing data are clearly apparent and will facilitate data entry later in the review process. Ideally, study assessment and data extraction should be conducted by two independent reviewers, and a third reviewer can be used if consensus cannot be reached on any discrepancies following discussion. It may be necessary to contact the original authors of a particular study if the data required for the systematic review are not presented in the published article but it is apparent that they were captured as part of the research study.

Typically, key data extracted from the primary studies include bibliographic details, descriptions of the setting, study population, details of the form and delivery of any intervention under investigation, outcome measures that were used, and results. Any factors that could also potentially affect the validity of the study results should also be documented. Although the general structure of all data extraction forms will be similar, it is often necessary to adapt forms to specifically meet the focus of a particular review. It is also sensible to pilot the extraction form on several studies.

For the prescribing error systematic review, data were extracted from the primary studies focusing on the setting (year and country, study duration, hospital setting), study methods (sampling of prescriptions and review processes, profession of data collector, means of detecting prescribing errors), prescribing error definitions used, error rates (including the nature of the error denominator, e.g., medication orders, patient admissions), severity of errors, type of errors, and medications commonly associated with errors. This allowed detailed consideration within the systematic review of the influence of variations in these factors on the reported error rates.

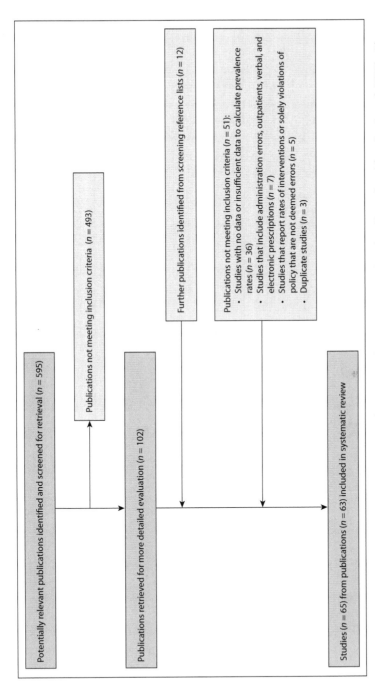

Figure 8.1 Study flow diagram of the screening process for the prescribing error systematic review. (Reproduced from Lewis et al. (2009) with permission from Wolters Kluwer Health/Adis. © Adis Data Information BV (2009). All rights reserved.)

Quality assessment

The relevant studies identified often vary greatly in quality. Different studies may examine the same issue in different ways, or examine different, but related, issues. Studies that present their data consistently and use high standards of reporting are clearly better suited for systematic reviews than studies with incomplete details concerning methodology, missing data, or inconsistent outcome measurement. As with any other step in the review process, assessing study quality should aim for the same high standards. In other words, the process should be explicit, documented, reliable, and free from bias. Indeed, the Preferred Reporting Items for Systematic reviews and Meta-Analyses (PRISMA) statement specifically requires that systematic reviews assess and report on the quality of the studies and data on which they are based (Liberati et al. 2009).

The most common way to assess study quality is to use one of the many published "critical appraisal checklists." However, it is important to recognize that:

- different checklists exist for different study designs, and most are designed for RCTs
- some checklists assign a score or grade to a study, whereas others simply remind the reviewer of key study characteristics
- even the best checklist does not eliminate the need to make judgments about the quality of a study.

Many different quality assessment checklists have been published in the literature, which often vary considerably in dimensions covered and complexity. Two of the most commonly applied instruments are the Jadad scale (Jadad et al. 1998) for use with RCTs and the Newcastle–Ottawa Scale (NOS) which was developed to assess the quality of nonrandomized studies (www.ohri.ca/programs/clinical_epidemiology/nosgen.pdf). The Jadad scale assesses clinical trials against three main criteria, namely whether (1) the trial is described as randomized, (2) the trial is described as double-blinded, and (3) a description of withdrawals and dropouts is provided. In contrast, the NOS uses a "star system" in which a study is judged on three broad perspectives: (1) the selection of the study groups; (2) the comparability of the groups; and (3) the ascertainment of either the exposure or outcome of interest for case–control or cohort studies, respectively.

In the absence of an appropriate checklist, study characteristics can be summarized qualitatively, including details about study design, methods of participant selection, and presence of potential bias in the conduct of the study or outcome measurement. Once complete, a quality assessment can be done to exclude studies from the systematic review that fall below predefined minimum standards for relevance or validity. Alternatively, it may be possible to grade or rank the component studies by quality, thereby giving greater emphasis to those studies of the highest quality.

Data synthesis

Once the data have been extracted and their quality and validity assessed, the findings from the individual studies may be combined or pooled to produce a summary outcome or "bottom line" on the acceptability, feasibility, or effectiveness of the intervention or activity under investigation. This stage of the process is often referred to as "evidence synthesis" and the type of evidence synthesis is chosen to fit the type of data within the review. For example, if a systematic review examines qualitative data then thematic analysis or meta-ethnography may be most appropriate (see Dixon-Woods et al. [2005] for a detailed critique of methods for synthesizing qualitative and quantitative evidence).

Close inspection of the data may indicate that it is too heterogeneous, or too sparse, to permit sensible pooling. Under these circumstances, a qualitative summary would be appropriate. In contrast, if homogeneous quantitative data are available these could be summarized statistically using meta-analysis techniques. Meta-analysis combines the quantitative outcomes of an intervention taking account of measures of variability both within and between studies. Different statistical approaches have been proposed to combine the numerical data, such as fixed-effects and random-effects models, which can be implemented in a range of bespoke or general statistical software packages (such as Revman or STATA software). The detailed methods of meta-analysis are beyond the scope of this chapter but can be found in several reference sources (see Higgins and Green [2008], or Borenstein et al. [2009], for further details).

Tests for heterogeneity are commonly used to decide on methods for combining studies in meta-analyses and for concluding consistency or inconsistency of findings (Higgins et al. 2003). Reports of meta-analyses commonly present a statistical test of heterogeneity that aims to determine whether there are genuine differences between the results of different studies (heterogeneity), or whether the variation in findings is compatible with chance alone (homogeneity). If heterogeneity is detected, it is important to examine possible factors that may explain apparent differences in the findings of the studies, such as the characteristics of the patients and settings, differences in measurement of outcomes, or differences in the nature or delivery of interventions.

Reporting

The findings from a systematic review should be summarized concisely and clearly, and put into context to inform the decision-making process. In particular, this should include issues such as the quality and heterogeneity of the included studies, the likely impact of bias, and reflections on the generalizability of the results so that a balanced summary of the usefulness of the review can be made.

Ideally, the format and content of a systematic review should follow the recommended guidelines for the type of study. The PRISMA statement has been developed to guide reporting of systematic reviews and meta-analyses of studies that evaluate healthcare interventions (see www.prisma-statement.org; Liberati et al. 2009). It consists of a 27-item checklist that includes items considered essential for the transparent reporting of systematic reviews. The statement also includes examples of good reporting for each checklist item, and references to relevant empirical studies and methodological literature explaining the rationale behind the use of the items.

Most of the PRISMA checklist items are also relevant for reporting systematic reviews of nonrandomized studies assessing the benefits and harms of interventions/exposures, but when reporting on systematic reviews of epidemiological studies it is necessary to consider additional items. In particular, the MOOSE (Meta-analysis Of Observational Studies in Epidemiology) guidance should be followed for reviews involving observational study designs (Stroup et al. 2000). The PRISMA and MOOSE guidelines have been developed to ensure that the information required for replication and verification of the systematic review is presented in a clear, organized, and accessible manner. Many of the leading academic journals now require evidence of adherence to such guidance before publishing systematic reviews.

Identifying and appraising systematic reviews

High-quality systematic reviews are regularly published in many of the leading medical journals and electronic databases, and structured queries have been developed to help locate systematic reviews and meta-analyses indexed with the PubMed database. In addition, many organizations now routinely publish electronic publications of systematic reviews on their own websites providing speedy access to frequently updated summaries of the available evidence, examples of which are listed in Box 8.3.

However, not all published systematic reviews are conducted or reported in a rigorous and unbiased manner. As with any research study, it is important to assess the strengths and limitations of a systematic review to help decide whether the findings should be applied in practice. Several organizations, such as the Centre for Evidence Based Medicine (see www.cebm.net) and the Critical Appraisals Skills Programme (CASP – www.phru.nhs.uk) in the UK have produced useful guidance to support the critical appraisal of systematic

> **Box 8.4 Useful websites for locating systematic reviews**
>
> - Agency for Healthcare Research and Quality (AHRQ): www.ahrq.gov/clinic/techix.htm
> - Canadian Agency for Drugs and Technologies in Health (CADTH): www.cadth.ca/index.php/en/hta
> - The Cochrane Library: www.cochrane.org
> - The Campbell Collaboration: www.campbellcollaboration.org
> - The Evidence for Policy and Practice Information and Co-ordinating Centre (EPPI-Centre): http://eppi.ioe.ac.uk/cms
> - The NHS Centre for Reviews and Dissemination: www.york.ac.uk/inst/crd
> - PubMed Clinical Queries: Find Systematic Reviews: www.ncbi.nlm.nih.gov/entrez/query/static/clinical.shtml#reviews

reviews. In line with these resources, the following questions serve as a framework:

- Is the research question well defined in terms of the population studied, the intervention under investigation, and the outcomes that were assessed?
- Was the search for studies comprehensive? Was the search strategy described? Was there screening of reference lists as well as electronic databases? Were non-English language studies included?
- Were the inclusion criteria to select studies appropriate?
- Was the quality of the included studies assessed? Was this undertaken by independent reviewers? Were the findings related to study quality?
- If the results of studies have been combined, was it reasonable to do so? Are the results similar from study to study? Were there tests for possible heterogeneity? Are the reasons for any variations in the results discussed?
- What are the main findings? How large is the result and how precise is it? How would the "bottom line" result of the review be summarized? Should pharmaceutical practice or policy change as a result of this review?

Summary and conclusions

Systematic review is a process that involves systematically identifying, critically appraising, and synthesizing the results of primary research studies. Reviews conducted using this approach aim to define the boundaries of what is and what is not known about a particular topic, and thereby inform decisions about the organization and delivery of healthcare services. They are prepared using strategies that limit bias in the selection and synthesis of research evidence, and are rapidly replacing traditional narrative reviews as a way of summarizing research evidence.

Systematic reviews are forms of original research designed to answer specific questions and require careful planning and execution. However, not all systematic reviews are of high quality, and it is important to be able to critically assess their applicability and validity. This chapter has provided an overview on how high-quality summaries of the research evidence relating to pharmaceutical practice and policy can be undertaken and critically appraised. Although the process described in the chapter is specific to systematic reviews, the procedures and techniques can be beneficial in a literature review process to understand the existing knowledge base in pharmaceutical practice and policy research.

Review questions/topics

1 What are the key differences between systematic reviews and traditional narrative reviews?
2 What are the most common types of selection bias in literature reviews?
3 Outline the five key steps in undertaking a systematic review.
4 What is a meta-analysis?
5 What key questions should be considered when critically appraising systematic reviews?

References

Borenstein M, Hedges LV, Higgins JPT, Rothstein HR (2009). *Introduction to Meta-analysis*. Chichester: Wiley.

Dixon-Woods M, Agarwal S, Jones D, Young B, Sutton A (2005). Synthesizing qualitative and quantitative evidence: a review of possible methods. *J Health Serv Res Policy* 10(1): 45–53.

Dwan K, Altman DG, Arnaiz JA, *et al.* (2008). Systematic review of the empirical evidence of study publication bias and outcome reporting bias. *PLoS One* 3(8): e3081.

Higgins JPT, Green S, eds (2008). *Cochrane Handbook for Systematic Reviews of Interventions*, Version 5.0.1 (updated September 2008). The Cochrane Collaboration. Available at: www.cochrane-handbook.org (accessed May 14, 2010).

Higgins JP, Thompson SG, Deeks JJ, Altman DG (2003). Measuring inconsistency in meta-analyses. *BMJ* 327: 557–60.

Jadad AR, Moher D, Klassen TP (1998). Guidelines for reading and interpreting systematic reviews: II. How did the authors find the studies and assess their quality? *Arch Pediatr Adolesc* 152: 812–17.

Lewis PJ, Dornan T, Taylor D, Tully MP, Wass V, Ashcroft DM (2009). Prevalence, incidence and nature of prescribing errors in hospital inpatients: A systematic review. *Drug Saf* 32: 379–89.

Liberati A, Altman DG, Tetzlaff J, *et al.* (2009). The PRISMA statement for reporting systematic reviews and meta-analyses of studies that evaluate health care interventions: explanation and elaboration. *Ann Intern Med* 151: W65–94.

Mathieu S, Boutron I, Moher D, Altman DG, Ravaud P (2009). Comparison of registered and published primary outcomes in randomized controlled trials. *JAMA* 302: 977–84.

Stroup DF, Berlin JA, Morton SC, *et al.* (2000). Meta-analysis Of Observational Studies in Epidemiology (MOOSE): A proposal for reporting. *JAMA* 283: 2008–12.

Suarez-Almazor ME, Belseck E, Homik J, Dorgan M, Ramos-Remus C (2000). Identifying clinical trials in the medical literature with electronic databases: MEDLINE alone is not enough. *Control Clin Trials* 21: 476–87.

Online resources

Agency for Healthcare Research and Quality (2007). *Methods Reference Guide for Effectiveness and Comparative Effectiveness Reviews*, Version 1.0 (Draft posted Oct 2007). Rockville, MD: AHRQ. Available at: http://effectivehealthcare.ahrq.gov/repFiles/2007_10DraftMethodsGuide.pdf.

Higgins JPT, Green S, eds (2008). *Cochrane Handbook for Systematic Reviews of Interventions*, Version 5.0.1 (updated September 2008). The Cochrane Collaboration. Available at: www.cochrane-handbook.org.

Khan KS, Riet G, Glanville J, Sowden AJ, Kleijnen J, eds (2001). *Undertaking Systematic Reviews of Research on Effectiveness: CRD's guidance for those carrying out or commissioning reviews.* York: University of York: CRD Report Number 4, 2nd edn. Available at: www.york.ac.uk/inst/crd/pdf/crdreport4_complete.pdf.

Public Health Resource Unit. 10 questions to help you make sense of reviews. Critical Appraisal Skills Programme (CASP). Available at: www.phru.nhs.uk/Doc_Links/S.Reviews%20Appraisal%20Tool.pdf.

9

Data collection methods

Richard R Cline

Chapter objectives

- To describe qualitative data collection methods
- To present quantitative data collection methods
- To introduce mixed data collection methods
- To discuss the strengths and limitations of specific procedures

Introduction

This chapter reviews data collection methods commonly employed in research on pharmaceutical practice and policy. After conceptualizing the research problem or question and conducting a review of the extant literature relevant to this research area, it is necessary to select a research method for the study, which may include some form of primary data collection (i.e., systematically accumulating data that do not already exist). This chapter reviews methods for data collection most often employed for pharmaceutical practice and policy research.

The chapter starts with a discussion of qualitative data collection methods and the philosophical foundations upon which these methods are based. Next, quantitative data collection methods are reviewed. The final section of this chapter introduces mixed data collection methods, a set of study designs integrating the qualitative and quantitative traditions that has recently become popular. The strengths and limitations of specific procedures within each of these traditions are also discussed.

Qualitative data collection methods

Qualitative data collection methods place significant emphasis on the individual or small group and their experience(s) with the phenomenon of interest in a study. The role of the investigator and his or her influence on the data collected (as an instrument for data collection and interpretation) are explicitly recognized with these techniques (Carpenter and Suto 2008). Researchers using these techniques stress the subjective experience of participants, conducting these studies within naturalistic settings, and describing a program or policy from the perspective of those involved with or affected by it. These methods emphasize the inductive approach, and serve to aid the researcher in developing hypotheses, models and tentative theories (Creswell 1994; Taylor and Bogdan 1998). As a result, these methods are often employed when there is little prior data bearing on the subject and when existing theory does not seem applicable to the problem. Qualitative data collection methods typically are identified with a constructivist theoretical orientation toward research (Table 9.1).

A note on qualitative sampling techniques

Qualitative data collection often employs purposeful sampling techniques such as theoretical sampling and the use of key informants (Carpenter and Suto 2008). In theoretical sampling the researcher begins interviewing and after the initial interviews starts to vary somewhat the type of individual interviewed (Taylor and Bogdan 1998). Interviews are continued until they yield no new information, or are saturated. The sample size for a qualitative study is rarely known beforehand for these reasons. In key informant interviewing, the researcher seeks out people who are intimately knowledgeable about the topic of interest and attempts to recruit them into the research project. As individuals of this type are often limited in number and/or difficult to access, it may not be possible to achieve saturation with key informants. Austin and Gregory (2006) illustrate the use of key informants in their study of the application of traditional tenure and promotion guidelines to clinical faculty members in Canadian colleges of pharmacy. These researchers used two types of key informants (clinical faculty members and academic administrators) as sources of data. In this study, key informants provided interviews, helped the researchers to analyze the documents relevant to the tenure and promotion process, and helped in clarifying ambiguous tenure and promotion policy documents.

Personal interviews

Personal interviews take place in a one-on-one setting, between the researcher and research participant. They are useful in situations in which the participant or process of interest is not directly observable (Taylor and Bogdan 1998). In the personal interview technique, researcher and participant meet in a private or semi-private location and begin discussing the topic of interest.

Table 9.1 Philosophical foundations of data collection methods

Data collection method	Theoretical perspective	Ontology (i.e., What is the nature of reality?)	Epistemology (i.e., What is the nature of the relationship between the researcher and the phenomenon being studied?)	Axiology (i.e., What is the role of values in research?)	Data collection method example
Qualitative	Constructivism	Reality is a subjective phenomenon; multiple realities exist	Researcher and phenomenon of interest interact in important ways that are not separable	The researcher's values will always influence the types of data collected and its interpretation	Semi-structured interview
Quantitative	Post-positivism	Reality is an objective phenomenon; only one exists	Researcher and phenomenon of interest are separable	The influence of the researcher's values can be minimized by adhering to accepted methodological standards and through peer review	
	Content analysis				
Mixed methods	Pragmatism	There is a true reality, but it is entwined with our social interactions with it	Researcher collects data by whatever means is best suited to address research question	Research is influenced by values to some extent, but use of multiple methods (triangulation) can help limit this	Triangulation design

Note: adapted from Creswell (1994), Creswell and Plano-Clark (2007), Greene (2007), Hunt (1991), and Johnson and Onwuegbuzie (2004).

The researcher may take notes as the interview proceeds, or may use audio recording to free him- or herself from the cognitive demands of note taking while remaining engaged with the participant. Many authors (Taylor and Bogdan 1998; Carpenter and Suto 2008) also recommend that the investigator record observations (including the general appearance of the informant, body language) after the interview concludes.

Two primary types of personal interviews have been distinguished (Flick 1998; Taylor and Bogdan 1998). In the unstructured or in-depth interview (Carpenter and Suto 2008), the investigator poses a very general question, or questions, to a participant who has experience with the phenomenon of interest. For example, if someone were interested in pharmacist–patient relationships, he or she might start by posing a question such as "During a typical week, what type of interactions do you have with patients who frequent the pharmacy where you practice?" The process is allowed to proceed organically "from the ground up," ending when the researcher or informant feels that there is no more information to share (Low 2007). In the semi-structured interview, an interview schedule or question route is used, which helps to guide the interview to semi-specific areas or topics in which the researcher is interested. As this type of interview employs a question route, it also is known as the focused interview. These topics may be generated from earlier unstructured interviews or earlier quantitative results that require more explanation. For example, the researcher may ask: "Since your enrollment in the prescription drug coverage program, how has your life changed?" This question might then be followed by a series of open-ended questions about other aspects of the program.

Among the advantages of the personal interview is its adaptability to a time and place that is agreeable just to the informant and the interviewer, and the fact that the personal interview is confidential, facilitating discussion of topics that may be perceived as socially stigmatized. Some disadvantages of the technique are that: it can be time-consuming to access and recruit a sufficient number of informants to achieve saturation, developing sound interview skills requires some practice, and accounts provided by informants are their interpretations and memories of their actions.

Focus groups

Focus groups involve the interviewing of small groups of participants by the researcher (Taylor and Bogdan 1998; Green 2007). These groups are composed of small numbers of informants, generally six to eight at a time. Although marketing researchers often recommend groups as large as 12 (Churchill 1995), other methodologists advise that focus groups should never contain more than eight because they can become difficult to control (Krueger and Casey 2000). With very small numbers of informants (two or three), group members may feel inhibited due the lack of confidentiality and because it is much easier for one individual to dominate the discussion (Churchill 1995). With groups larger than eight, all group members may lack time to adequately share their views. In addition, very large groups promote the tendency to free ride or remain silent, relying on other group members to share their experiences, thoughts, and feelings. Focus groups generally last from 1 hour to 2 hours (Carpenter and Suto 2007).

The informants who make up a focus group are chosen based on their shared experience of some product, program, or other phenomenon, or perhaps on some shared characteristic(s) (e.g., community dwelling women aged 45–60 or pharmacists practicing in institutional settings). This homogeneity of group composition is important for two reasons: first, for analysis purposes it becomes very difficult to compare the experiences of two or more groups if these group members are combined into a single group; and, second, participants are more likely to feel comfortable and share experiences within a group with similar participants.

As with the semi-structured interview, focus groups employ a questioning route designed to focus the inquiry and to keep the discussion from ranging too far afield from the topic of interest (Krueger and Casey 2000). As the researcher must attempt to solicit input from multiple informants, they are usually referred to as a moderator, as opposed to an interviewer, in this context. Each group is audio- or video-recorded so that verbatim transcriptions or detailed notes can be made and analyzed. In addition, a second member of the research team records notes at each focus group, recording such things as descriptions of the group members and other details of group dynamics that may add context (Green 2007).

Among the advantages of focus groups are that group members may be prompted to remember and share thoughts and opinions by simply hearing those shared by other members (Taylor and Bogdan 1998). In addition, they can be more cost-effective because the researcher is soliciting information from many people at the same time, potentially increasing the reliability of the data (Churchill 1995). However, the investigator must still conduct the number of focus groups needed to reach saturation (often a minimum of three) (Krueger and Casey 2000). The disadvantages of this method include the fact that the researcher must find an appropriate space and conduct the focus groups at a time that is convenient for the majority of the target population. Sensitive topics (e.g., prescription drug diversion) are poor subjects to study with focus groups and are better addressed with personal interviews.

Unstructured observation methods

Observation methods involve the researcher as observer, recorder, and interpreter of research participants' actions, conversations, body language, etc., in a setting relevant to the investigation (Flick 1998). Observing subjects engaged in an activity can serve as a useful method for describing and understanding the structure and process used in an environment (Neuman 2006). As observations are made in a naturalistic setting these techniques are generally referred to as field research.

In nonparticipant observation, the researcher remains detached from the research participants (Caldwell and Atwal 2005). The researcher may remain hidden or use video-taping. The goal is to make observations without altering the natural environment where the activity occurs. For example, Bissell et al. (2000) used nonparticipant observation as one data collection procedure in their study of the appropriateness of advice given to consumers with regard to nonprescription medicines use in the UK. One member of the research team unobtrusively took field notes on such factors as whether the pharmacy appeared to be busy and whether a pharmacist or medicines counter assistant provided the advice. These observations were then used to supplement audio-recording transcripts of each advice-giving encounter. In participant observation, the researcher becomes known to the participants in the study (Flick 1998). The researcher's level of involvement may range from being known to the research participants but remaining apart from them, to actually becoming one of the participants in the group under study, so that the emotions and attitudes of the group can be experienced (Mack et al. 2005; Neuman 2006). For example, Sinclair and coworkers (2008) used participant observation to augment focus groups in their study of Canadian oncologists who made use of a decision-making tool to prioritize funding for several new cancer drugs being evaluated for inclusion in a drug formulary. Research team members sat with each group of oncologists while prioritization occurred, taking field notes (the traditional data recording method used in these techniques) on nonverbal communication among members. Audio transcripts were analyzed and augmented with these notes. Oncologist behaviors, such as burying their faces in their hands, provided insight into the arduous nature of these decisions which might not have been apparent with verbal transcripts alone.

Advantages of observation methods include the researcher actually witnessing the actions of research participants and their accompanying dynamics. In addition, they can provide insight into subjects that research participants are unwilling or find it difficult to talk about (Green 2007). In addition, research participants may be unaware of the cultural knowledge that they use and possess in their daily activities because it is embedded in their daily lives. Disadvantages of the technique include the time required to complete observational studies (which can range from 6 months to as long as 2 years) and the difficulty involved in recording everything that one sees, hears, smells, etc., in field notes (Mack et al. 2005). Finally, several authors (Mack et al. 2005; Green 2007) caution against the belief that, as the researcher is in close proximity to the situation, the data generated are more valid than other types of qualitative data. Steps can be taken to increase validity (such as video-recording or employing multiple observers), but the fact that the researcher is the primary data collection instrument makes the activity inherently subjective.

Quantitative data collection methods

Quantitative data collection methods emphasize collecting objective data on prespecified variables of interest from individuals who are members of the population of interest in a research project (Neuman 2006). The investigator serves as a neutral third party who attempts to minimize the effect of personal biases on the data collected (Broom and Willis 2007). Quantitative researchers stress collecting structured data in a reliable manner using data collection instruments, such as survey forms, that require observations to fit within prearranged categories and ranges (Creswell 1994). These methods emphasize the deductive approach, and serve to aid the researcher in testing hypotheses and verifying (actually, attempting to falsify) models and theories (Godfrey-Smith 2003). These methods are often employed when the subject of a research problem is well structured, when there are at least some prior data, or when existing theory seems applicable. Quantitative data collection methods typically are identified with a post-positivist theoretical orientation toward research (Table 9.1).

Structured observation

In structured observation methods, the researcher seeks out and records data from situations that are of interest in an investigation (Mateo 1999). Structured observations focus the investigator's attention on specific variables and employ data collection forms requiring the researcher to record observations using specific categories or values. Structured observation techniques can be classified as either direct or indirect. Direct observation methods entail observing or recording the situation under study as it occurs and may be either disguised or undisguised. When using a disguised observation method, the researcher attempts to blend in with other individuals in the situation of interest when collecting data. Alternatively, hidden video- or audio-recording may be used to gather data. In undisguised observation the investigator identifies him- or herself before starting the observation. The advantage of disguised observation is that research participants are less likely to alter their behaviors in response to being observed (Churchill 1995). In contrast, undisguised observation permits the researcher to question those involved in the situation of interest about any number of other, unobservable variables of interest. Schommer and Wiederholt (1995) used direct disguised observation in a natural setting (community pharmacies) in their study of pharmacist, patient, and environmental variables affecting the occurrence, length, and content of pharmacist–patient communication. Patients were made aware that they were observed only after their interactions with pharmacists, at which time structured personal interviews were used to collect other data.

Indirect observation techniques (sometimes referred to as unobtrusive measures) focus on the evidence left behind after the activity occurs (Boyd et al. 1985). For example, Mott and Kreling (1998) audited recently processed written prescription orders from 10 community pharmacies, recording data on variables such as drug name, quantity, whether a proprietary or generic product was dispensed, and the type of prescription drug insurance. The advantage of indirect methods is that they provide the most assurance that the researcher cannot be detected and the behavior of interest can be studied in its purest form. The clear disadvantage is the difficulty in devising good unobtrusive measures (Neuman 2006).

Self-reports

Self-report data collection methods are all obtrusive because they involve confronting the research participant directly and gathering specific information (Kerlinger 1973; Neuman 2006). The key feature that differentiates the self-report methods described below is their method of administration. The validity of all self-reports is, at best, subject to the limits of memory.

The structured face-to-face interview involves the researcher approaching the potential research participant, posing the required questions, and recording the answers provided on data collection forms. The emphasis in the face-to-face interview is obtaining standardized answers to the questions presented. This can be difficult for at least two reasons (Fowler 1995): first, terms employed in questions may be understood differently from person to person; and, second, interviewer training must be standardized so that interviewers understand what types of clarification are permissible in these situations. If this is not done, the reliability and validity of the data collected are threatened.

Face-to-face structured self-reports have several advantages. They are effective in encouraging study participation because the social interaction between investigator and participant builds rapport (Frankel and Wallen 1996). Also, those administering the questionnaire can help clarify items on the survey form, thus reducing item nonresponse. The interviewer may also prompt the respondent when incomplete responses are given. Finally, this form of administration may be the most feasible when conducting research within populations with low literacy levels. The disadvantages of this method include its high cost and the fact that social desirability bias is likely to be prevalent. Computer-assisted personal interviewing (CAPI), in which the interviewer uses a laptop computer to record answers, seems to moderate at least some of these disadvantages. In their study of the National Longitudinal

Study of Youth (NLSY), Baker et al. (1995) found that interviews completed using CAPI were on average 20 percent shorter than those completed using paper and pencil, and that respondents were more likely to respond to questions on sensitive topics.

Telephone interviews provide an attractive method of collecting self-reports in some projects. Several authors suggest that the percentage of homes with some type of telephone in the USA is approximately 97 percent, providing very good coverage of most of the population (Blumberg and Luke 2008). Techniques such as random digit dialing (RDD) can be used efficiently to generate random samples of any required size. However, the increasing prevalence of cell phone-only households (17.5 percent in 2008) may be beginning to limit the utility of this technique, suggesting that researchers must use dialing procedures that ensure coverage of this group.

The advantages of telephone interviews include their relative cost-effectiveness and the speed with which they can be conducted compared with in-person interviews (Churchill 1995). Telephone interviews retain several advantages of face-to-face interviews including the ability to prompt the respondent for complete answers and to clarify vague terms. They may also be better at eliciting truthful responses to sensitive questions. Among the disadvantages of this method are its slightly lower response rates compared with face-to-face interviewing and the fact that interview length is typically limited (about 25 minutes) (Frankel and Wallen 1996; Neuman 2006).

Structured self-report data are also commonly collected by mail survey (Churchill 1995). This method typically involves selecting a sample of participants and mailing the questionnaire to these individuals with (at a minimum) a cover letter containing a number of necessary elements (Dillman 2000) and a postage paid return envelope. The reader is cautioned that collecting self-report data via mail questionnaire may seem simple, but is difficult to do well. Thus, the interested reader should refer to Chapter 10, as well as the references contained herein, before beginning this process.

Advantages of the mail survey include the fact that it is generally among the lowest costs of the four methods discussed in this section (Neuman 2006). If necessary, this type of data collection can be conducted by a single researcher. Mail surveys are self-administered, increasing anonymity. Thus, respondents are more likely to respond truthfully to sensitive questions. The disadvantages of the mail questionnaire include its limited usefulness in populations with low literacy and in groups who are homeless. Finally, response rates to mail surveys can be among the lowest of those discussed here. For example, a review of studies using mailed self-report forms, which appeared in medical journals in 1991, found that the mean response rate among physicians was 54 percent, whereas among patients it was 60 percent (Asch et al. 1997). However, close adherence to techniques known to increase response rates (e.g., multiple contacts, financial incentives) can boost response rates to 70–80 percent (Dillman 2000).

The advent of widespread broadband internet availability in US households (approximately 63 percent in 2009) (Anonymous 2004; Horrigan 2009) provides a new mode for self-report data collection in the form of the web-based survey. Although a number of popular web-based applications exist for designing and implementing internet surveys, many (if not most) of the principles governing mail data collection form design and administration still apply in this new delivery mode (Dillman 2000). The advantages of the internet survey as a method of self-report collection include its low cost, the fact that prompts can be used for incomplete or unanswered items, and the ability to provide context-sensitive instructions for terms that may be unclear. As with mail surveys, these web surveys are often superior to in-person and telephone surveys at eliciting truthful responses to controversial questions. This method of data collection has numerous disadvantages. First, it requires widespread written and computer literacy within the target population. Second, access to high-speed internet coverage is unequal across various demographic groups, with older adults, those living in rural areas, and those with lower incomes being all less likely to reside in a home with a broadband connection (Horrigan 2009). Finally, a recent meta-analysis of studies comparing response rates between internet and mail surveys suggests that rates are approximately 10 percent higher in traditional mail surveys (Shih and Fan 2008).

Content analysis

Content analysis is an unobtrusive data collection technique used when the researcher seeks to describe the content of some form of communication (written, verbal, or visual) (Weber 1990; Holdford 2008). It is a means of collecting structured data from various types of messages. Content analysis is used for three general categories of research questions (Frankel and Wallen 1996). It is useful for developing descriptive data from a communication medium (e.g., who, what, where). It is also useful in determining the impact of some social phenomena. Thus, a researcher might seek to better understand changes in Medicare beneficiaries' attitudes toward Medicare by content analyzing newspaper stories written about the program both before and after the Medicare Part D benefit was implemented. Finally, content analysis may be used in studies designed to test hypotheses. Using this approach, Sleath et al. (1997) used transcriptions from audio-recordings of patient–physician encounters to study the factors influencing patient-initiated prescribing of psychotropic medications. They then tested a set of hypotheses about variables that they believed were likely to influence this activity.

When using content analysis to collect data the researcher must first decide upon indicators of the construct(s) of interest (Kerlinger 1973;

Holdford 2008). After indicators are selected, the analyst must decide on the coding scheme. Neuman (2006) suggests that four basic dimensions of the selected indicators can be coded: frequency (the number of times something occurs), direction (e.g., Do facial expressions appear to be happy or sad?), intensity (the strength of an indicator in a given direction), and space, or the size of the message, sometimes coded as column inches or minutes. The reliability of the investigator's coding is a central concern when collecting data in content analysis (Holdford 2008). To increase reliability, content analysts generally employ a minimum of two coders who analyze the same units of analysis. The degree of correspondence between the coders, or interrater reliability, can then be estimated using a variety of statistical indices (Perrault and Leigh 1989). Low reliability suggests that the coding scheme developed for the project requires revision or that coders require further training.

Content analysis has several potential advantages as a data collection method (Weber 1990). First, communication is a pervasive activity, so many areas of pharmaceutical practice and policy research are amenable to it. Second, it is an unobtrusive data collection method, limiting the potential for the observation process itself to confound the phenomenon of interest (Neuman 2006). Its limitations include the fact that its use is restricted to descriptive and correlational research designs. Content analysis studies cannot demonstrate a causal relationship between two variables.

Clinical/biophysiological data

The collection of clinical data may be of use in the evaluation of various pharmaceutical practices and policies (Lepper and Titler 1999). For example, in a study of a pharmacist intervention to decrease tobacco use, a proximal outcome of interest would be changes in tobacco use that occur among program participants. Observational methods are likely to be burdensome and self-reports will likely be subject to social desirability bias. One alternative, clinical and biophysiological data collection, is to check for saliva or blood levels of cotinine, a nicotine metabolite, among study participants (Centers for Disease Control and Prevention 2002).

The advantages to using clinical and biophysiological measures in data collection include the fact that most are highly reliable, if carefully conducted, and valid, if chosen carefully (i.e., the researcher must be sure to carefully select clinical markers that are considered good measures for the outcome of interest) (Polit and Tatano-Beck 2003). For example, cotinine is produced only from nicotine exposure. However, a high level may indicate that patients are exposed to a large amount of secondary cigarette smoke, and not that they themselves use tobacco. The primary disadvantages to this data collection method are the fact that they are often more invasive than other methods discussed above and can be costly if the equipment needed must be purchased (Lepper and Titler 1999).

Mixed methods data collection

Mixed methods data collection involves integrating qualitative and quantitative data collection methods in a study to develop a more complete understanding of the pharmaceutical practice and policy that one seeks to evaluate (Creswell and Plano Clark 2007). It should be noted that integration is a key term used in this brief definition, implying that one method informs the other, producing an understanding not possible if either had been used alone. This approach to research is referred to as triangulation (Tritter 2007). Mixed methods designs involve the mixing of various data collection techniques described above. As such, this section introduces no new methods as such, but instead focuses on the design types used in mixed methods research. Qualitative data collection methods are often identified with a pragmatic theoretical orientation toward research (see Table 9.1).

Mixed methods research designs

Creswell and Plano Clark (2007) propose that mixed methods studies, and the various frameworks that methodologists have advanced for the classification of their research designs, can be classified into just four main types of study design. The following discussion draws heavily on their classification work, and the interested reader is encouraged to refer to their text for a more complete exposition on this topic.

The purpose of the triangulation design is to obtain complementary data on the same topic. It is known as a single-phase design because the different types of data are collected concurrently. Equal weighting is given to each type of data. For example, quantitative and qualitative data may be collected simultaneously and then interpreted simultaneously in the research report. Alternatively, the researcher may collect quantitative data using a structured survey while collecting qualitative data using open-ended, semi-structured survey items addressing similar topics. The qualitative responses might then be used to interpret quantitative results.

The primary advantages to this type of design are, first, that it makes intuitive sense to most researchers. The triangulation design is efficient, because both types of data are collected at approximately the same time. Second, each type of data can be collected separately by a member of the research team who is expert in the area. The disadvantages of triangulation designs are that the amount of effort and expertise required to conduct this type of study are considerable (because equal weight is accorded to each type of data, research teams composed of investigators expert in both types of data collection must be formed). Also, results from the two data types may not agree, necessitating the collection of further data to resolve the problem.

The embedded design is used when one type of data will play a supportive role in the research project. It is useful when the investigator needs to insert a smaller quantitative (or qualitative) research component within a larger qualitative (or quantitative) study to answer a different question that could not be answered with a single type of data. This design might use a one- or a two-phase design (i.e., a design in which one type of data collection is followed by another). For example, one variant of the embedded design might be to conduct a qualitative study to determine what types of stimuli are likely to be effective in an experimental manipulation and then to design an experiment(s) using these findings.

The advantages of the embedded design include its adaptability to the needs of the research project. Also, the design typically is simpler for graduate students, since one method requires less data than the other. This design may also be more appealing to funders, because the emphasis is often placed on the quantitative part of the study. Its disadvantages include the challenges in specifying the purpose of collecting qualitative (quantitative) data within a quantitative (qualitative) study and the difficulty in integrating the results when the methods are used to answer two different research questions.

The explanatory design is a two-phase design with the intent of using qualitative data to help explain initial quantitative findings. Thus, qualitative data collection follows the main, quantitative portion of the study. The two variations on the explanatory mixed methods design are the follow-up explanations model wherein the researcher requires qualitative data to explain results obtained during the quantitative phase of the investigation and the participant selection model wherein the researcher uses quantitative data to purposefully identify and sample participants for further qualitative study. Here, the qualitative data usually become the primary focus of the research.

As the explanatory design is a two-phase design, it can be used by a single researcher collecting one type of data at a time. This family of designs works well for both multiphase and single-phase mixed methods investigations. Finally, this variety of designs is appealing to quantitatively oriented researchers because it usually begins with a quantitative worldview. However, explanatory designs have drawbacks, including the amount of time required to implement the phases in sequence and the design decisions that the investigator must make (e.g., using the same or different participants for each study phase).

In the exploratory design, the results from a first phase (typically qualitative) are used to develop a second phase. The exploratory design is useful when theory is lacking in the area of interest or important variables are unknown. Here, the qualitative data often receive greater emphasis. In one variation of this design, the taxonomy development model, the qualitative data collection phase helps uncover variables with the quantitative phase used to test the relationships among them. This design variant was employed by Cline and Gupta (2006) who first collected data on the types of features older adults believed were most important in a government-provided prescription benefit program using several focus groups. Thematic analysis of the focus group transcripts revealed five attributes that were consistently mentioned. These attributes were incorporated into a number of hypothetical drug plan designs that a separate group of older adults rated on a structured survey form. Analysis using multiple regression revealed the most important attributes in this group.

As a result of its similarities to the explanatory design, the advantages of the exploratory design are analogous. The disadvantages of this design include the time required to conduct studies using exploratory designs, the difficulties that sometimes arise when trying to specify a research plan to an institutional review board when design decisions must be made based on the first phase of the study, and making design decisions regarding the use of the same versus different study participants in each phase.

Summary and conclusions

This chapter has introduced three varieties of data collection procedures commonly employed by pharmaceutical practice and policy researchers. Qualitative methods are often employed when the purpose of a study is inductive or exploratory. These techniques include in-depth interviews, focus groups, and participant observation techniques. They are especially valuable when there are few prior data in an area and when theory is lacking. Qualitative data collection methods are especially useful for generating details from the perspective of the participant that describe the structure and process of a phenomenon.

Quantitative data collection methods are used when the purpose of a study is deductive. These methods include structured observation methods, content analysis, and structured self-reports, such as telephone and mail surveys. Quantitative data collection techniques require that the granularity of data be reduced so that they can be recorded on data collection forms with preordained categories and values. Although these procedures are sometimes used in exploratory studies, they are most often associated with hypothesis testing or model falsification.

Mixed data collection methods are often used when the researcher believes that neither qualitative nor quantitative data collection techniques alone are likely to answer the research question(s) posed in a study. These studies involve integrating both qualitative and quantitative methods in an attempt to develop a greater understanding of a phenomenon than would be possible otherwise. The goal in a mixed methods study is to select data collection methods with complementary strengths and overlapping weaknesses.

In conclusion, a wide variety of primary data collection techniques has been developed by social scientists. Before selecting a data collection method, the researcher must first define the research problem carefully, decide on the variety of data (qualitative, quantitative, or mixed) most likely to be useful in addressing the question, and finally review the advantages and disadvantages of the individual methods before selecting a data collection procedure to be used.

These data collection procedures are powerful and, when applied rigorously and combined with appropriate analysis techniques, can yield answers to many questions. However, students of these techniques should keep two points in mind: first, the data collection methods discussed here are just methods – they are not the subject of science but its servants; and, second, no data collection technique, no matter how elegantly applied, can compensate for a poorly conceived research project, i.e., the first job of any scientist is to understand the important questions to ask in his or her chosen field.

Review questions/topics

1 How are decisions about the data collection techniques to be employed in a study linked to the research question(s) that must be addressed?
2 What are the advantages of qualitative data collection methods compared with quantitative data collection approaches?
3 What are the advantages of quantitative data collection methods compared with qualitative data collection techniques?
4 Compare and contrast direct (obtrusive) structured data collection methods with indirect (unobtrusive) data collection methods.
5 What facets of a given research problem would you consider when attempting to decide among mixed methods research designs?

References

Anonymous (2004). *A Nation Online: Entering the broadband age.* Washington, DC: US Department of Commerce.

Asch DA, Jedrziewski KM, Christakis NA (1997). Response rates to mail surveys published in medical journals. *J Clin Epidemiol* 50: 1129–36.

Austin Z, Gregory PAM (2006). Promotion and tenure: clinical faculty at schools of pharmacy in Canada. *Pharm Educ* 6: 267–74.

Baker RP, Bradburn NM, Johnson RA (1995). Computer-assisted personal interviewing: an experimental evaluation of data quality and cost. *J Off Stat* 11: 413–31.

Bissell P, Ward NR, Noyce PR (2000). Appropriateness measurement: application to advice-giving in community pharmacies. *Soc Sci Med* 51: 343–59.

Blumberg SJ, Luke JV (2008). Wireless substitution: Early release of estimates from the National Health Interview Survey, January–June 2008. National Center for Health Statistics. Available from: http://www.cdc.gov/nchs/nhis.htm. December 17, 2008.

Boyd HW, Westfall R, Stasch SF (1985). *Marketing Research: Text and cases*. Homewood, IL: Richard Irwin.

Broom A, Willis E (2007). Competing paradigms in health research. In: Saks M, Allsop J (eds), *Researching Health: Qualitative, quantitative, and mixed methods approaches*. Thousand Oaks, CA: Sage, 16–31.

Caldwell A, Atwal A (2005). Non-participant observation: using video tapes to collect data in nursing research. *Nurse Researcher* 13(2): 42–52.

Carpenter C, Suto M (2008). *Qualitative Research for Occupational and Physical Therapists: A practical guide*. Ames, IA: Blackwell Publishing.

Centers for Disease Control and Prevention (2005). *Third National Report on Human Exposure to Environmental Chemicals*. Atlanta, GA: Department of Health and Human Services.

Churchill GA (1995). *Marketing Research: Methodological foundations*. Fort Worth, TX: The Dryden Press.

Cline RR, Gupta K (2006). Drug benefit decisions among older adults: a social judgment analysis. *Med Decis Making* 26: 273–81.

Creswell JW (1994). *Research Design: Qualitative and quantitative approaches*. Thousand Oaks, CA: Sage.

Creswell JW, Plano Clark V (2007). *Designing and Conducting Mixed Methods Research*. Thousand Oaks, CA: Sage.

Dillman DA (2000). *Mail and Internet Surveys: The tailored design method*. New York: John Wiley & Sons.

Flick U (1998). *An Introduction to Qualitative Research*. Thousand Oaks, CA: Sage.

Fowler FJ (1995). *Improving Survey Questions*. Thousand Oaks, CA: Sage.

Frankel JR, Wallen NE (1996). *How to Design and Evaluate Research in Education*. New York: McGraw-Hill.

Godfrey-Smith P (2003). *Theory and Reality*. Chicago: University of Chicago Press.

Green J (2007). The use of focus groups in research into health. In: Saks M, Allsop J (eds), *Researching Health: Qualitative, quantitative, and mixed methods approaches*. Thousand Oaks, CA: Sage, 112–32.

Greene J (2007). *Mixed Methods in Social Inquiry*. San Francisco, CA: Wiley.

Holdford D (2008). Content analysis methods for conducting research in social and administrative pharmacy. *Res Soc Admin Pharm* 4: 173–81.

Horrigan J (2009). Home broadband adoption. Available at: www.pewinternet.org/Reports/2009/10-Home-Broadband-Adoption-2009.aspx.(accessed June 23, 2009).

Hunt S (1991). *Modern Marketing Theory*. Cincinnati, IL: South Western.

Johnson RB, Onwuegbuzie AJ (2004). Mixed methods research: a research paradigm whose time has come. *Educ Res* 33(7): 14–26.

Kerlinger FN (1973). *Foundations of Behavioral Research*, 2nd edn. New York: Holt, Rinehart & Winston.

Krueger RA, Casey MA (2000). *Focus Groups: A practical guide for applied research*, 3rd edn. Thousand Oaks, CA: Sage.

Lepper HS, Titler MG (1999). Program evaluation. In: *Using and Conducting Nursing Research in the Clinical Setting*. Philadelphia, PA: WB Saunders Co., 90–104, 256–67.

Low J (2007). Unstructured interviews and health research. In: Saks M, Allsop J (eds), *Researching Health: Qualitative, quantitative, and mixed methods approaches*. Thousand Oaks, CA: Sage, 74–91.

Mack N, Woodsong C, MacQueen KM, Guest G, Namey E (2005). *Qualitative Research Methods: A data collector's field guide*. Research Triangle Park, NC: Family Health International.

Mateo MA (1999). Psychosocial measurement. In: *Using and Conducting Nursing Research in The Clinical Setting*. Philadelphia, PA: WB Saunders Co., 256–67.

Mott DA, Kreling DH (1998). The association of prescription drug insurance type with the cost of dispensed drugs. *Inquiry* 35: 23–35.

Neuman WL (2006). *Social Research Methods: Qualitative and quantitative approaches*. Boston, MA: Allyn & Bacon.

Perrault WD, Leigh LE (1989). Reliability of nominal data based on qualitative judgments. *J Market Res* 26: 135–48.

Polit DF, Tatano-Beck C (2003). *Nursing Research: Principles and methods*. Philadelphia, PA; Lippincott, Williams & Wilkins.

Schommer JC, Wiederholt JB (1995). A field investigation of participant and environment effects on pharmacist-patient communication in community pharmacies. *Med Care* 33: 567–84.

Shih T, Fan X (2008). Comparing response rates from web and mail surveys: a meta-analysis. *Field Methods* 20: 249–71.

Sinclair S, Hagen NA, Chambers C, Manns B, Simon A, Browman GP (2008). Accounting for reasonableness: exploring the personal internal framework affecting decisions about cancer drug funding. *Health Policy* 86: 381–90.

Sleath B, Svarstad B, Roter D (1997). Physician vs. patient initiation of psychotropic prescribing in primary care settings: A content analysis of audiotapes. *Soc Sci Med* 44: 541–8.

Taylor SJ, Bogdan R (1998). *Introduction to Qualitative Research Methods: A guidebook and resource*, 3rd edn. New York: Wiley.

Tritter J (2007). Mixed methods and multidisciplinary research in health care. In: Saks M, Allsop J (eds), *Researching Health: Qualitative, quantitative, and mixed methods approaches*. Thousand Oaks, CA: Sage, 299–318.

Weber RP (1990). *Basic Content Analysis*, 2nd edn (Sage University Paper Series on Quantitative Applications in the Social Science #07-049). Newbury Park, CA; Sage.

Online resources

Journal of Mixed Methods Research (Sage Publications). Available at: www.sagepub.com.

Sites Related to Survey Research. Available at: www.srl.uic.edu/srllink/srllink.htm.

Sydenstricker-Neto J. *Research Design and Mixed Method Approach: A hands-on approach*. Available at: www.socialresearchmethods.net/tutorial/Sydenstricker/bolsa.html.

10

Survey design

Marcia M Worley

Chapter objectives

- To discuss the appropriateness of using surveys
- To review survey development
- To describe survey mode selection
- To describe survey implementation
- To discuss data preparation

Introduction

Survey research is widely used in the domain of pharmaceutical practice and policy research. According to Doyle (2008): "Surveys can be a powerful and useful tool for collecting data on human characteristics, attitudes, thoughts and behaviors." For example, a survey can be used to collect data related to: (1) patient satisfaction with medication therapy management services, (2) patient attitudes and beliefs about the Medicare Part D prescription drug program, (3) practitioner views on direct-to-consumer advertising, and (4) pharmacists' views on their roles as medication therapy management providers.

This chapter examines principles of survey design in order to provide a practical framework and resources for designing effective surveys. The first section of the chapter starts with a discussion about the appropriateness of using surveys. Next, the topics of survey development, including writing survey questions, types of response scales, and formatting the survey, are reviewed. The second part of the chapter describes issues related to survey mode selection and mail survey implementation, including a discussion of errors common to survey research. Lastly, data preparation issues related to coding and cleaning are discussed.

Getting started in the survey research process

When do you use a survey?

Beginning researchers often make the common mistake of starting first with the idea that they want to conduct a survey, even before research questions have been formulated. If research questions are clearly thought out and defined, then the research project can be properly designed to answer such questions. The next question that the researcher should ask is: "Are my research questions best answered by using a survey, or is another data collection method appropriate?" A survey can be one option for collecting the data if a researcher needs to collect primary data (data that do not already exist). Although surveys can be difficult, costly and time-consuming to develop and implement correctly, they can yield rich primary data.

Using existing instruments versus instrument development

The first step in the survey development process is for the researcher to examine the literature for existing survey instruments that may be usable, which can save the researcher valuable time in the research process. Surveys in the literature can sometimes be used exactly as written, or can be modified to apply to the researcher's individual research project. If this approach is used, it is important to assess the reliability and validity of these items in the current study. Question modifications and using questions in different samples can affect the psychometric properties of the questions (Meadows 2003). It is important to obtain permission to use the survey, which may include copyright issues, from the originator of the survey. Appropriate acknowledgment through referencing should be followed (Meadows 2003).

If new survey questions need to be developed, there are numerous avenues to explore for developing such questions. The researcher can develop new questions by putting pen to paper. Also, focus groups and interviews can be used to generate survey questions as discussed in Chapter 9.

Constructing the survey

Writing survey questions

If existing surveys cannot be found in the literature to meet the researcher's needs, survey questions must be written. Writing survey questions is a challenging and complex endeavor, part art and part science. This step of the survey design process is one in which the researcher should plan on spending considerable time and effort, because this is a critical step to ensure that reliable and valid data are collected to answer the research questions.

Open-ended versus close-ended questions

When preparing survey questions, the researcher can choose to use questions that have an open-ended or a close-ended response option. An open-ended question allows the respondent to write in any answer to the question. A close-ended question includes response options that are provided by the researcher, thereby "forcing" the respondent to choose from a set of predetermined response options. For example, a researcher could ask a patient an open-ended question: "Please describe who is involved in your daily diabetes medication management."

A close-ended question could be: "Who is involved in helping you manage your daily diabetes medication?" (Check all that apply):

- Doctor
- Pharmacist
- Nurse
- Family member or friend
- Other.

There is a tradeoff between the control of using a close-ended question versus the variety of responses and the richness of the data that can be obtained using an open-ended question. In addition, data analysis is different for open-ended questions (typically qualitative data) than for close-ended questions (quantitative data). (The reader is encouraged to investigate resources for appropriate data analysis.) Often a researcher will use a combination of both closed and open-ended questions in a survey, depending on the type of information that they are collecting (Fowler 1995, 2009; Passmore et al. 2002; Meadows 2003).

Question components

Close-ended survey questions or items can be conceptualized as containing a question "stem" and a "response format" (Passmore et al. 2002). Simply put, the "stem" is the statement responded to or the question to which the respondent answers. The response format structures the person's response. Guidelines to follow for writing effective question stems that can produce high-quality data include writing stems that are relatively short (<20 words), clear, and simple (Passmore et al. 2002). Also a good rule to follow is not to use modifiers in the stem, e.g., "usually," which can cause confusion. Additional guidelines for writing effective question stems include writing stems that do not influence respondents to answer in a particular way, are not socially or culturally offensive, and focus on one specific variable (Fowler 1995; Passmore et al. 2002).

Additional question writing tips

To increase the quality of data collected, the researcher is encouraged to use simple language and avoid the use of technical terminology, abbreviations, and jargon. Avoiding words that are ambiguous or nonspecific, or have multiple meanings is also encouraged. The researcher should avoid using double-barreled questions, e.g., "How do you feel about your medications and the pharmacist who is helping you with your medications?" In this example, the respondent may have negative feelings about their medications and positive feelings about their pharmacist, or vice versa, so they would not be able to accurately answer the question (Oppenheim 1992; McColl et al. 2001; Meadows 2003). A better approach would be to ask this question as two separate questions. The researcher should word survey questions in such a way that they can be answered by the person in the sample with the lowest education level (Barker 1994). The researcher can use tools such as the FOG Index (Gunning 1952) or the Flesch Reading Ease formula and the Flesch–Kincaid Grade Level formula (Klare 1974–75) to assess the reading level of the survey questions.

Common scales used in survey research

When close-ended questions are used to measure variables, the respondent can be asked to choose from an array of response categories, choose a response by writing in a number corresponding to a value on a particular scale, or circle a response on a scale. If scales are used as response formats, there are various forms.

Likert scale

The Likert scale is one of the most commonly used response scales in pharmaceutical practice and policy research (Likert 1932). It is a respondent-friendly scale typically used to measure attitudes, beliefs, and opinions. Characteristically, a declarative statement is presented with a Likert scale containing response options representing degrees of agreement and disagreement. For example, a respondent can be asked to respond to the following statement by choosing a response option from the 5-point accompanying scale:

"My pharmacist will alert my physician if there are problems with the medications that I am taking:" 1 = strongly disagree; 2 = disagree; 3 = neutral; 4 = agree; 5 = strongly agree.

A 7-point scale can also be used, providing more response options and allowing for increased variability in responses. An example of a 7-point scale is:

1 = very strongly disagree; 2 = strongly disagree; 3 = disagree; 4 = uncertain; 5 = agree; 6 = strongly agree; 7 = very strongly agree.

In some cases, the researcher may not want to allow the respondent to have the response option of "neutral" or "uncertain" and wants to "force" the respondent to choose either a positive or a negative response option (Sapsford 2007).

Semantic differential scale

The semantic differential scale has its roots in the attitude research of Osgood and colleagues (1957). A semantic differential scale can be used by the researcher to assess a respondent's attitudes and opinions about an object, person, product, or service. A semantic differential scale comprises pairs of adjectives or phrases that are opposite in meaning. The respondent must then choose the adjective or phrase that best describes their attitude or opinion. These adjectives or phrases can be bipolar or unipolar, depending on the study purpose. An example of a bipolar adjective pair, expressing opposite attributes, would be trustworthy and deceitful. An example of a unipolar adjective pair, focusing on the absence and presence of one attribute, would be trustworthy and not trustworthy (DeVellis 2003). For example, patients can be asked to rate their opinions about pharmacists conducting medication therapy management services using the following adjective pairings: inexpensive–expensive; valuable–not valuable.

Visual analog scale

A visual analog scale is a visual representation of the semantic differential scale. The scale is set up so that a line is drawn between the two adjectives or phrases that represent a continuum. The respondent is asked to place a mark on the line that represents their perception or viewpoint, and then the researcher measures the distance from one side of the line (DeVellis 2003; Sapsford 2007).

Rank order scale

A rank order scale can be used if the researcher wants the respondent to rank or order preferences related to a certain variable. For example, a survey question that asks a person enrolled in the Medicare Part D prescription drug plan:

> Please indicate which of the following aspects were most important to you when you chose your Medicare Part D prescription drug plan. Please indicate importance using the following items 1 (most important) to 5 (least important) (Passmore et al. 2002):

- The amount I pay each time I pick up a prescription (copay)
- Remaining with my current pharmacy
- The plan formulary covers most of the drugs that I take

- The amount the plan charges in monthly premiums
- The plan covered prescription costs in the "doughnut hole" (coverage gap).

It is important to keep the list relatively short in an effort to decrease respondent burden. An issue that arises with this type of scale is that the elements of a rank order list cannot be compared as degrees of magnitude.

Additional scales

A Guttman scale is a hierarchical scale in which a respondent ranks items such that respondent agreement (or affirmative response) with an item indicates the respondent's agreement with all preceding items in the scale (DeVellis 2003; Rattray and Jones 2007). If the researcher uses a Guttman scale, they are focused on the point where the respondent changes their answer from an affirmative answer to a negative response (DeVellis 2003). A Thurstone scale is similar to a Guttman scale, and is developed such that individual items in an item pool have different levels of intensity of the attribute being investigated (DeVellis 2003). The scale items are separated by equal intervals, and an "agree–disagree" response format is used. The researcher then examines the level of agreement or affirmative responses to the items (DeVellis 2003). The difficulty in developing Guttman and Thurstone scales correctly often limits their use in pharmaceutical practice and policy research. Novice researchers are encouraged to consult appropriate references when designing either of these scales.

Ordering of survey questions

Organization of survey questions according to survey topics is one way to present questions in a logical order that will not be confusing to the respondent. This approach decreases respondent burden and increases the likelihood that respondents will put thought into their answers. As Schwarz (1996) stated: "A questionnaire is like a conversation which typically evolves in accordance with societal norms." According to Dillman (2007), the order in which the topics appear in the survey should be from most to least important to the respondent. Criteria for the first question is that it should be able to be answered by everyone, easy to answer, interesting, and shows the respondent that the first question is connected or related to the study purpose which has been outlined in the cover letter (Dillman 2007).

Questions that are used to collect information on personal or sensitive topics (e.g., household income) should be placed near the end of the survey. The logic behind this placement is that, by the time the respondent reaches this question, he or she will have already invested time in the survey, is most likely to realize the importance of the survey, and is interested in the survey topic. Similarly, questions related to demographics (e.g., gender, marital status) should be placed at the end of the survey because typically they are not viewed by respondents as important (Dillman 2007).

Formatting the survey

Evidence in the literature suggests that respondent-friendly surveys can improve survey response rate, thereby minimizing nonresponse to survey questions (Dillman et al. 1993). Elements of a respondent-friendly survey include ease of completion, avoiding confusion when completing the survey, and evoking positive or neutral affective responses from the respondent, compared with a negative response to the survey (Dillman et al. 1993). In the earlier work by Dillman and Reynolds (1990), and Dillman et al. (1991), these researchers empirically identified elements of surveys that made them *not* respondent friendly. These included multistep folding of paper, inclusion of multiple inserts in the questionnaire (making the task of survey completion appear complicated), instructions that were lengthy, complicated, and contradictory, and respondent confusion of where to begin the survey due to use of multiple graphics.

An effective survey layout in which clear instructions and graphics are used to guide the respondent through the survey will reduce the chance that questions or entire sections of the survey will be missed by the respondent (Dillman 2007). When deciding on the survey format, using a booklet format is the preferred method (Dillman 2007). This format is commonly used in survey research. Typically, respondents can navigate through them and respond to questions more easily and with less error, compared with unconventional formats. Examples of unconventional formats include printing text on both sides of a piece of paper and using a staple in the upper left corner to hold the pages together, or folding the paper in an unusual manner that may make the survey appear complicated to the respondent (Dillman 2007). The goal when formatting a survey is to keep it simple, yet professional. The respondent should not have to spend time trying to figure out the survey format (e.g., multiple folds) at the same time as he or she is trying to answer the survey questions (Dillman 2007). Dillman (2007) provides other examples and pictures of effective survey formats.

Visual appearance of the survey

The researcher should spend time focusing on the visual appearance of the survey, because this can influence survey response rate (Meadows 2003; Dillman 2007). One aspect of visual appearance is survey length. Researchers are often concerned with this, often believing that the shorter the survey the better. Respondent burden can lead to decreased quality of item responses and to lower response rates. However, evidence in the literature supports the tenet that, if the topic is of interest to the respondent, then using a longer survey will not affect response rates (Oppenheim 1992; McColl et al. 2001; Dillman 2007).

The question and the response format can be laid out vertically or horizontally. In a vertical presentation, the response options are listed on separate lines as a list below the question. With a horizontal presentation, the response options are listed below the question going from left to right across the page. Typically, a vertical presentation format is preferred because it separates the question from the response option and facilitates data entry (Bourque and Fielder 1995; Meadows 2003). In addition, vertical formats can increase visual appeal because the survey appears neater (Sudman and Bradburn 1982; Dillman 2007).

The survey should look professional, neat, and organized. This can be accomplished by judicious use of graphic designs that are professional in nature. It is important to include sufficient "white space" in the survey booklet so that the survey does not appear crowded and disorganized. Text should be written in a standard typeset and font, typically 12 point. To facilitate ease of reading, it is recommended that font size be not less than 10 point (Meadows 2003). If a researcher is administering a survey to a sample of older people or to people who are visually impaired, the researcher should consider increasing the font size (e.g., 14–16 point), as well as the amount of white space in the survey, to make the survey more user friendly (Meadows 2003). Use of colored paper for survey printing can be considered, with the choice depending on the sample and the topic being studied (Meadows 2003).

Design considerations for modes of survey administration

There are a variety of modes to administer a survey, including self-administration, interview, telephone, mail, and electronic. As discussed in Chapter 9, each of theses modes has advantages and disadvantages. The focus of this section is on highlighting the survey design issues that should be considered for different modes of survey administration.

Self-administration

The questions must be worded and the survey designed with clear instructions so that respondents can easily and accurately complete the survey on their own. Designing unambiguous questions and response options, using the appropriate white space, font, and layout, are also critical when a survey is self-administered. The researcher must carefully consider the cognitive ability (e.g., reading level) of respondents when using the self-administration mode. For example, if the reading level of the target population is low, the researcher may choose to administer the survey using an interview, rather than self-administration. As surveys administered via mail and electronic means are self-administered, they have similar design considerations (Meadows 2003; Fowler 2009).

Telephone survey

If telephone interviewing is used to administer a survey, the researcher must consider several issues related to questionnaire design and types of data that can be collected. Visual aids, including pictures and diagrams, used to help respondents answer questions in self-administered or interview-administered surveys, cannot be used when a survey is done over the telephone. The researcher should also pay attention to questions that are sensitive and/or personal in nature, because it may be more difficult to elicit a truthful response from a respondent over the telephone, compared with an anonymous, self-administered survey (Meadows 2003; Fowler 2009).

Interview survey

Face-to-face survey administration, also known as personal interviewing, has many design advantages compared with self-administered surveys, and to some extent telephone interviewing. As the interviewer is in close proximity to the respondent, a well-trained interviewer can read respondent cues, e.g., confusion or lack of understanding of a particular question. Interviewers can clarify respondent questions and survey instructions, and ask probing questions to clarify responses. To some extent, it can be easier for the interviewer to establish a positive rapport with the respondent, compared with self-administered survey procedures. This can be helpful to the interviewer if collection of sensitive and/or personal data is necessary, as well as the collection of data using longer survey questionnaires (Meadows 2003; Fowler 2009).

Considerations for selecting modes of administration

In selecting the survey mode, consideration should be given to a variety of issues before the survey is developed, because these issues have an impact on survey design. The researcher needs to think about the purpose of the study when making decisions about the survey mode. For example, if a researcher is interested in studying homeless teenagers' medication use experiences, a mail or telephone survey would not help the researcher meet the survey objectives. Interviewing these individuals would be a better approach. However, if a researcher wanted to assess the satisfaction of people with medication therapy management services at a pharmacy clinic, a telephone- or mail-administered survey could be appropriate (Meadows 2003).

The researcher's access to resources is also a paramount consideration. Resources include the costs and time related to developing and implementing the survey, as well as analyzing survey data (e.g., personnel costs). For example, if a researcher is using a telephone-administered survey, the researcher must spend time and money to develop and administer the survey, with often extensive costs associated with training personnel and then paying them to administer the telephone survey (Meadows 2003).

The researcher must also pay careful attention to access to and character-istics of the target population. For example, if a researcher is interested in collecting older people's opinions about the Medicare Part D prescription drug benefit at a national level, then interviewing individuals should probably not be the mode of survey administration. In this example, accessibility to the poten-tial respondents would most likely be through a mail survey (Meadows 2003).

How the survey is administered can impact many other things in the survey research process. Data quality, survey-related error, researcher bias, and response rates can each be affected depending on how the survey is adminis-tered (Meadows 2003). For example, a researcher can introduce biases into the data collection process as they administer a survey over the telephone, thereby affecting the quality of the data collected.

Survey implementation

Review by an expert panel, pre-testing, pilot testing

After the survey questions have been developed and the survey formatted, the next step for the researcher is to conduct small-scale tests of the survey. The first step is to have experts in one's discipline review the survey – often called review by a panel of experts. At this stage, content experts provide valuable feedback on survey content. The panel of experts can also provide comments about the survey format and ease of completing the survey (DeVellis 2003).

The next steps in the survey design process are to conduct a pre-test of the survey and then pilot test the survey. These two distinct survey tests are conducted before full sample survey administration in an effort to work out any "bugs" that the survey may contain. A pre-test involves administering the survey to a small group of individuals ($n = 10$–30) who are similar to the target population for whom the researcher wants to generalize study results. Pre-test results can help the researcher identify problems with clarity of ques-tions, response categories, directions, and other problems that may interfere with respondents completing the survey consistently and accurately. A pilot test has a similar purpose to a pre-test, but, as it is typically larger in size ($n = 100$–300), the study instrument can be psychometrically tested (reliabil-ity and validity analyses), statistical analyses conducted, and survey imple-mentation procedures tested. The researcher must consider the time, money, and personnel resources that will be needed to conduct these tests. Although it is advisable to conduct both a pre-test and a pilot test, the above-mentioned constraints have to be considered (Passmore et al. 2002; Meadows 2003; Czaja and Blair 2005).

Mail survey implementation

Dillman's (2007) tailored design survey method is considered by many to be the gold standard of mail survey implementation methods. In brief, tailored design is a process in which the researcher works to increase respondent trust and rewards (or perceptions of rewards), and decrease respondent costs associated with responding to the survey. This is achieved by focusing on elements of survey design and implementation issues that decrease survey-related sources of error. For example, the total design method advocates using multiple contacts with potential respondents, creating respondent-friendly surveys, using first-class postage stamps on survey return envelopes, personalization of all survey contact materials, and use of a token financial incentive (Dillman 2007). As this is just a brief overview of the tailored design method, it is advisable to consult resources that cover this topic in depth (Dillman 2007).

Elements of the total design method can be used in the survey implementation process. Dillman (2007) advocates the use of five contacts with potential mail survey respondents as a technique to maximize mail survey response rate. Based on principles of the social exchange theory (Thibaut and Kelley 1959; Homans 1961; Blau 1964; Dillman 2007), the purpose of multiple respondent contacts is to increase response rate. Therefore, each contact contains a different message appealing to the respondent to complete the survey. The first contact is a pre-notification letter, which is typically sent 1 week before the arrival of the first survey. This first contact to the respondent is meant to introduce the survey topic to the potential respondent and emphasize that the individual's response would be appreciated.

The second mail contact is the survey packet, which is mailed to potential respondents 1 week after the pre-notification letter. The survey packet contains a cover letter, token incentive, survey, and return envelope. The purpose of the cover letter is to explain to the respondent in more detail (compared with the pre-notification letter) what the purpose of the survey is and to emphasize the importance of the respondent's individual response. A token incentive, which can be financial in nature, is typically included with the cover letter. Based on the principles of the social exchange theory (Thibaut and Kelley 1959; Homans 1961; Blau 1964), financial incentives ranging from US$1 to US$5 can be used, and have been shown to have a positive impact on response rates (Dillman 2007). The postage-paid return envelope is the third component of the survey packet. Using real postage stamps instead of a business reply envelope (containing postage) shows that the researcher is sending something of monetary value to the respondent, which the respondent could potentially use for another purpose. This can invoke a sense of obligation on the respondent's behalf to complete and return the survey in the postage paid envelope.

The third mail contact is a follow-up reminder postcard which is mailed to respondents 1 week after the first survey packet has been sent. The follow-up postcard serves as a reminder for respondents and encourages them to respond if they have not already done so. Approximately 1–2 weeks after the follow-up reminder postcard has been mailed, the researcher should send a second survey packet to those respondents who have not completed and returned the first survey. This fourth contact does not include an incentive, but contains all the other elements of the first survey packet. The difference is that the cover letter has a stronger tone and appeal for response, compared with the first cover letter. Finally, a fifth contact can be done in which nonrespondents are contacted by telephone or certified mail procedures. Although the repeated contacts survey implementation procedure has been presented in the context of self-administered mail surveys, these approaches could be modified to apply to different modes of survey administration, e.g., electronic surveys.

Reducing survey error

Each mode of survey administration discussed in this chapter is prone to a variety of errors. Although these errors can never be completely eliminated, the goal in survey research is to minimize these errors as much as possible. Minimization of errors in survey research can be accomplished by various design and implementation strategies.

Sampling and measurement error

Sampling error results from the fact that usually researchers collect data from a sample of respondents, and rarely the entire population. Random sampling can help to minimize this type of error (Dillman 2007). Measurement error is a result of respondents not answering survey questions with consistency and accuracy, which can lead to responses that are not comparable in a meaningful way with other respondents' answers (Dillman 2007). Reliability and validity analyses are two methods that can be used to assess measurement error in survey research. Measurement error can be reduced by spending time and effort on question development and survey formatting to create a respondent-friendly survey. Review of the survey by a panel of experts, pretesting, and pilot testing can all be used to help identify measurement issues, thereby minimizing this type of error (Dillman 2007).

Coverage error

Survey research is susceptible to coverage error, which results when there are errors in the sampling frame or list of the individuals to be sampled, such that each individual in the population does not have an equal chance of being sampled (Dillman 2007). For example, if a researcher is conducting a telephone-administered survey and the sampling frame is drawn from the telephone book, coverage error could result because the people who do not have a home/landline telephone will not be listed in the telephone book. These people have no chance of being included in the sampling frame, so coverage error becomes problematic. Researchers need to do their homework to be knowledgeable about how the sampling frame is generated. Having this key piece of information will help the researcher to assess the impact, if any, of coverage error. Coverage error can have a profound impact on the researcher's ability to generalize the survey results to the target population (Dillman 2007).

Nonresponse error

Nonresponse error occurs when the individuals who respond to a survey differ from those who do not respond to a survey, with respect to a characteristic(s) that is important in understanding the research question(s). Any time the researcher obtains a survey response rate of less than 100 percent, nonresponse error exists. This error becomes increasingly problematic in studies involving lower response rates (Dillman 2007). Creating a respondent-friendly questionnaire can help to reduce nonresponse error. Also, choosing an easy and interesting first question can help to "hook" the respondent into completing the survey, which can help minimize nonresponse error. Careful attention to survey implementation, as discussed previously with the concept of multiple respondent contacts, can have the greatest effect on increasing response rate, thereby reducing nonresponse error (Dillman 2007).

Preparing data for analysis

After data are collected, the data file must be prepared for data analysis so that the results can be generated. After surveys are returned from respondents, typically the date returned and an identification number are marked on the survey. The identification number will be used to enter the survey information into a database for analysis. If a second survey is to be sent to initial non-responders, the identification number will also serve as a means of identifying which individuals have and which have not responded, before a second survey packet is mailed (Williams 2003; West 2007).

A researcher must assess the surveys for completeness of survey responses in order to determine whether or not the survey can be used for data analysis. He or she should set guidelines for determining what constitutes a "useable" number of responses. This is determined based on the number of total questions in the survey, as well as the relative importance of the survey item to answering the research question(s). Researchers can use various methods of replacing missing data (e.g., means, modes, or predicting the missing value using a regression model) (Downey and King 1998). However, one must keep in mind that there is a balance between inputting a few versus many missing data points. Maintaining the integrity of the dataset must be at the forefront of a researcher's decision-making process.

After data entry into the database is complete, the next step in the data preparation process is referred to as data cleaning. During this process, data are examined to identify incorrect data entry. To clean the data, frequency counts can be run on all items to check for inaccuracies. For example, if a survey question used a response scale from 1 to 5, and a response of 7 was found upon running a frequency check for that particular question, the researcher would examine the survey to see where the error occurred and make the correction. Although time and personnel intensive, in some cases a double-data entry method of entering survey data into a database can be employed. After two individuals have completed data entry, the data entry is checked for points of incongruence. Although data cleaning can be a time-intensive process, it is well worth the time investment to ensure that the data entry process is as error free as possible and the researcher can be confident of the integrity of the dataset (Williams 2003; West 2007).

The process of converting survey data into data that can be statistically analyzed is called data coding (Williams 2003). If respondents are asked to choose a number from a response scale (e.g., the Likert scale) to answer a close-ended survey question, then the numeric value chosen by the respondent can be entered into the computer database. If respondents are asked to choose responses from categories, the categories must be assigned a number that can subsequently be recorded in the database. The researcher should consider how the data will be coded even at the survey design stage (Williams 2003; West 2007). An example of this concept would be for the researcher to assign numbers to the response categories associated with a survey question and place this directly on the survey. These numbers will subsequently be linked to variables in the database. After data are collected, the researcher can use these codes to enter data in a database for statistical analysis (Williams 2003; West 2007).

The preparation of a codebook is essential to accurate data record keeping and interpretation of results. A copy of the survey, either paper or electronic, can be used to develop the codebook by using the numeric values or codes assigned to each survey question. The codebook can serve as a reference for personnel who are responsible for data entry to ensure consistency of data entry, as well as the researcher during the data analysis and interpretation process. It is advisable for the researcher to consult a biostatistician when establishing the codebook, because these coding decisions are directly related to data analysis. In addition, it is important to test the usability of the codebook during the pre-test and pilot test phases to work out any problems before the full-scale study is conducted (Williams 2003; West 2007).

Summary and conclusions

This chapter has introduced important concepts for effective survey design and implementation. An understanding of these concepts is essential for researchers using surveys to collect data in the pharmaceutical practice and policy domain. Survey design focuses on the importance of question development, choice of response scales, and formatting the survey. Another important aspect of the survey research process is selecting how the survey will be administered, which is related to the survey design process. Finally, effective survey implementation and data preparation are needed to use the survey to collect and analyze the data necessary to answer the study research questions.

In conclusion, an astute researcher will spend the necessary time thinking about and defining research questions, and selecting the appropriate data collection method to answer these questions. If a survey is the method chosen by the researcher to answer the research questions, then he or she should remember this important take-home message: designing surveys is much more than writing questions, putting them on paper, and administering them to respondents. It is a process and a series of decisions that must be carefully planned and implemented. When the researcher invests the time on these aspects of the research process, it will pay multiple dividends in terms of the quality of data obtained and the scientific contributions to our discipline.

Review questions

1 What are some of the challenges associated with developing new survey questions compared with using existing survey questions?
2 What are the differences between open-ended and close-ended questions in terms of researcher control and the type of data collected?
3 What are the main attributes of the response scales discussed in the chapter? Can you think of examples that you have used or would like to use in your own research projects?
4 What are the main issues that the researcher must address when formatting a survey?
5 What are the different modes of survey administration and related survey design issues?
6 What types of error should you consider when you are designing and implementing a survey? How can you use design and implementation strategies to minimize these sources of error?
7 What are the steps that need to be completed by the researcher to prepare data for analysis?

References

Barker PJ (1994). Questionnaire. In: Cormack DFS, ed. *The Research Process in Nursing*, 2nd edn. London: Blackwell Scientific: 215–27.

Blau PM (1964). *Exchange and Power in Social Life*. New York: Wiley.

Bourque LB, Fielder EP (1995). *How to Conduct Self-administered and Mail Surveys*. Thousand Oaks, CA: Sage Publications, Inc.

Czaja R, Blair J (2005). *Designing Surveys: A guide to decisions and procedures*, 2nd edn. Thousand Oaks, CA: Sage Publications, Inc.

DeVellis RF (2003). *Scale Development: Theory and applications*, 2nd edn. Thousand Oaks, CA: Sage Publications, Inc.

Dillman DA (2007). *Mail and Internet Surveys: The tailored design method*, 2nd edn. Hoboken, NJ: Wiley.

Dillman DA, Reynolds, R (1990). *Reasons for Not Responding by Mail to the 1990 U.S. Census: Hypotheses for research*. Technical Report No. 90-103. Pullman, WA: Washington State University, Social & Economic Sciences Research Center.

Dillman DA, Reynolds RW, Rockwood TH (1991). *Focus Group Tests of Two Simplified Decennial Census Forms*. Technical Report No. 91-39. Pullman, WA: Washington State University, Social & Economic Sciences Research Center.

Dillman DA, Sinclair MD, Clark JR (1993). Effects of questionnaire length, respondent-friendly design, and a difficult question on response rates for occupant-addressed census mail surveys. *Public Opin Q* 57: 289–304.

Downey RG, King CV (1998). Missing data in Likert ratings: a comparison of replacement methods. *J Gen Psychol* 125: 175–91.

Doyle JK (2008). "Introduction to Survey Methodology and Design." Available at: www.sysurvey.com/tips/introduction_to_survey.htm (accessed October 15, 2008).

Fowler FJ (1995). *Improving Survey Questions: Design and evaluation*. Thousand Oaks, CA: Sage Publications.

Fowler FJ (2009). *Survey Research Methods*, 4th edn. Thousand Oaks, CA: Sage Publications.

Gunning R (1952). *The Technique of Clear Writing*. New York: McGraw Hill, 4.

Homans G (1961). *Social Behavior: Its elementary forms*. New York: Harcourt, Brace, & World.

Klare GR (1974–75). Assessing readability. *Reading Res Q* 10: 62–102.

Likert R (1932). A technique for the measurement of attitudes. *Arch Psychol* 140: 1–55.

McColl E, Jacoby A, Thomas L, *et al.* (2001). Design and use of questionnaires: A review of best practice applicable to surveys of health service staff and patients. *Health Technol Assess* 5(31): 1–256.

Meadows KA (2003). So you want to do research? 5: Questionnaire design. *Br J Commun Nursing* 8: 562–70.

Oppenheim AN (1992). *Questionnaire Design: Interviewing and attitude measurement*, 2nd edn. London: Pinter.

Osgood CE, Suci GJ, Tannenbaum PH (1957). *The Measurement of Meaning*. Chicago, IL: University of Illinois Press.

Passmore C, Dobbie AE, Parchman M, Tysinger J (2002). Guidelines for constructing a survey. *Fam Med* 34: 281–6.

Rattray J, Jones MC (2007). Essential elements of questionnaire design and development. *J Clin Nursing* 16: 234–43.

Sapsford R (2007). *Survey Research*, 2nd edn. Thousand Oaks, CA: Sage Publications, Inc.

Schwarz N (1996). *Cognition and Communication: Judgmental biases, research methods, and the logic of conversation*. Mahwah, NJ: Erlbaum.

Sudman S, Bradburn N (1982). *Asking Questions: A practical guide to questionnaire design*. San Francisco, CA: Jossey-Bass.

Thibaut JW, Kelley HH (1959). *The Social Psychology of Groups*. New York: Wiley.

West DS (2007). Survey research methods in pharmacy practice. In: *The Science and Practice of Pharmacotherapy II*. Lenexa, KS: American College of Clinical Pharmacy, 67–77.

Williams A (2003). How to … write and analyse a questionnaire. *J Orthodont* 30: 245–52.

Online resources

American Association for Public Opinion Research. Best Practices for Survey and Public Opinion Research. Available at: www.aapor.org/bestpractices.

Encyclopedia of Survey Research Methods. Sage Reference Online. Available at: www.sage-rereference.com/survey.

Other resources

Asarch A, Chiu A, Kimball AB, Dellavalle RP (2009). Survey research in dermatology: guidelines for success. *Dermatol Clin* 27: 121–31.

Murray P (1999). Fundamental issues in questionnaire design. *Accid Emerg Nursing* 7: 148–53.

Sprague S, Quigley L, Bhandari M (2009). Survey design in orthopedic surgery: getting surgeons to respond. *J Bone Joint Surg Am* 91: 27–34.

Sue VM, Ritter LA (2007). *Conducting Online Surveys*. Thousand Oaks, CA: Sage Publications, Inc.

11

Statistical analysis

John P Bentley

Chapter objectives

- To describe common statistical terminology
- To explain descriptive and inferential statistics
- To understand criteria for selecting statistical tests
- To describe the use of a variety of statistical techniques

Introduction

Scientists have access to many different sets of tools to help them make sense of the world; one of those is the statistics toolbox. All scientific disciplines have benefited from the application of statistical methods. The intent of this chapter is to provide the reader breadth of coverage, rather than depth, of the many different types of statistical techniques. Some terminology necessary to understand the framework used in this chapter is initially reviewed; however, no attempt is made to replace an introduction to statistics provided in numerous textbooks. The approach taken is to encourage the reader to consider the many options available and to begin to understand how decisions are made with respect to choosing a particular statistical technique.

The purpose and function of statistics in research

Most authors differentiate between two functions served by statistics: descriptive and inferential. Descriptive statistics is generally concerned with organizing, tabulating, summarizing, depicting, or describing data, essentially reducing a large quantity of data to a manageable (and useable) form. Examples include measures of central tendency, variability, and correlation. In general, the collection of data from all members of a given population (i.e., a census) is not plausible. Thus, scientists are interested in drawing valid conclusions about a population based on information contained in a sample from that population. This is the realm of inferential statistics and the primary focus of this chapter.

There are two general activities of statistical inference: estimation and hypothesis testing. Estimation is concerned with providing a specific value, the estimate, thought to be a good approximation of an unknown population parameter such as the population mean, proportion, difference in two population means, or regression coefficient (i.e., point estimation), and associating with this estimate a measure of its variability, usually taking the form of a confidence interval for the parameter of interest (i.e., interval estimation). The other aim of statistical inference, hypothesis testing, is concerned with aiding decision makers to draw a conclusion about a hypothesized value of an unknown parameter. As most introductory statistics books note, these two functions are really not all that different because one may use a confidence interval to arrive at the same conclusion reached through hypothesis testing.

Most of the techniques discussed in this chapter use both functions of inferential statistics, estimating population parameters and testing hypotheses, such as whether a given variable significantly improves the prediction of another variable while mathematically adjusting for a set of other variables. It is important to note that the methods associated with hypothesis/significance testing are not without controversy (Harlow et al. 1997; Nickerson 2000). Despite the controversy, hypothesis testing continues to be widely reported in the scientific literature.

Some important concepts

Types of research studies

Previous chapters have discussed different types of study designs including experimental and nonexperimental (sometimes called observational) designs. In an experimental study an experimenter is able to determine the levels (conditions) of at least one independent variable (IV) and has control over how participants are assigned to the levels, usually through some type of randomization process. Experimental studies with randomization are usually considered the gold standard with respect to establishing causal effects (although not all agree, e.g., Heckman 2005).

The importance of this distinction is not that it is a major determinant of the choice of statistical technique (Tabachnick and Fidell 2007). Indeed, most of the techniques discussed in this chapter can be applied to data from either experimental or observational studies (statistical software does not "know" whether a variable was manipulated). However, the confidence attached to a given result is usually associated with the study type.

Often it is difficult or even impossible to implement a randomized controlled experiment to study a causal effect of interest. This may be because of ethical issues or cost considerations, or because the findings of an experiment may be difficult to generalize to other populations or settings. Researchers in these situations may be required to use nonexperimental designs, yet may desire to go beyond descriptive statements about the association between two variables and make statements about cause–effect relationships. Both statisticians and econometricians have spent a great deal of effort attempting to devise methods for estimating causal effects from nonexperimental data. In essence, attempting to control or adjust for confounders, or variables, related to the IV of interest, and also causally related to the outcome. Confounders can (falsely) obscure or accentuate the relationship between the focal IV and the outcome. Techniques include those that attempt to control for variables that have been measured or observed (e.g., multivariable adjustment, see Katz [2003]; propensity scores, see Rubin [1997]) and those that attempt to also control for unmeasured confounders (e.g., fixed effects regression methods for longitudinal data – see Allison [2009]; instrumental variables estimation – see Grootendorst [2007]). A review of these methods is beyond the scope of this chapter (see Wunsch et al. [2006], Schneeweiss [2007], and Normand [2008] for more information), but it is important to recognize that the techniques discussed in this chapter are explicitly utilized in the implementation of these methods. For example: multivariable adjustment with a continuous outcome variable is a simple application of linear regression; propensity scores methods typically utilize the technique known as logistic regression; and instrumental variable estimation is often conducted with a variation of regression known as two-stage least squares regression.

Classification of variables

The choice of statistical technique is strongly influenced by the type of variables that one is studying, so it is important to consider methods used to classify variables. One common method for variable classification, discrete versus continuous, pertains to the values that can be assigned to a variable. A discrete variable is characterized by gaps in the values that it can assume. Examples include sex, the number of visits to an emergency room (sometimes called a count variable), and response to treatment measured as excellent, good, fair, or poor (sometimes called an ordinal variable). Some use the terms "qualitative" or "categorical" for a discrete variable. A continuous variable does not possess the gaps characteristic of a discrete variable and can take on any value within a defined range (Tabachnick and Fidell 2007). Examples include blood pressure, temperature, and height. A "quantitative" variable can be used to describe a continuous variable. Discrete variables can sometimes be treated as continuous variables when analyzing data. In addition, continuous variables can be collapsed into discrete variables by specifying cutoff values on the continuous scale. This can sometimes be useful when there are meaningful cut-off values, but in other situations it can lead to substantial loss of information.

Another related classification scheme addresses the level of measurement associated with a variable (i.e., nominal, ordinal, interval, ratio). This scheme is described in Chapter 4 and provides more information than the discrete–continuous dichotomy. However, there is substantial overlap. For example, a nominal variable must be discrete and interval variables are usually treated as continuous (although there are exceptions such as count data, which are best conceptualized as interval and discrete). The distinctions between and within these categorization schemes can be fuzzy (Norman and Streiner 2008), but nevertheless they serve a useful purpose in aiding the researcher in selecting an appropriate technique for data analysis.

A third method of variable classification is according to their conceptual relationships, i.e., the variable is conceptualized as an IV, dependent variable (DV), or control variable (sometimes called a covariate, extraneous variable, or nuisance variable). This classification depends on the context of the study, rather than some inherent mathematical structure of the variable (Kleinbaum et al. 2008). IVs can be set to a desired level (i.e., the treatment in an experimental design) or observed as they occur in a sample. Nominal variables that are IVs are called factors (especially in the context of experimental design) and the different categories of a factor are called levels. In any given study, there may be variables that affect relationships of other variables, but they generally are not the focal interest of the study. These variables are called control variables.

One other useful classification scheme differentiates between latent and manifest variables. Although there is some disagreement (Bollen 2002), perhaps the most straightforward definition of a latent variable is provided by Skrondal and Rabe-Hesketh (2004):

> A latent variable [is] a random variable whose realizations are hidden from us. This is in contrast to manifest variables where the realizations are observed.

This distinction is probably most widely recognized in the social sciences, most notably in psychology, where latent variables are thought of as hypothetical constructs measured by multiple observed (manifest) indicators (implemented through the techniques of confirmatory factor analysis and structural equation modeling). For example, self-esteem, personality, and quality of life are hypothetical constructs that are measured indirectly by using observed responses to a set of questions. However, latent variables are used in a wide variety of applications and disciplines (although often with different names and terminology) (see Skrondal and Rabe-Hesketh [2004] for an overview). A brief review of some of the techniques that utilize latent variables is presented in a subsequent section.

A classification of statistical techniques

Classification criteria

A number of authors have presented classification frameworks for statistical techniques (e.g., Hair et al. 2006; Tabachnick and Fidell 2007). Techniques are often grouped according to the research question addressed by the analysis. Although the questions in such frameworks can be numerous, a simple way of grouping techniques by research question is whether the technique can be classified as a dependence or an interdependence technique (Hair et al. 2006). Dependence techniques involve the prediction or explanation of a variable (or set of variables, i.e., DVs) by another variable or variables (i.e., IVs). Interdependence techniques do not define variables as being independent or dependent; rather these procedures examine all variables concurrently, often focusing on the underlying structure among either the variables or the entities (often individual participants) that produced the values on the variables. Although this is a useful distinction, aspects of dependence and interdependence can be combined in a single analysis.

In addition to the research question addressed by a technique, other information that is helpful in organizing statistical techniques includes: (1) the number and type of DVs (i.e., continuous vs discrete); (2) the number and type of IVs; (3) the consideration of IVs as control variables; (4) the inclusion of latent variables in the analysis; and (5) the presence of correlated data.

Description of statistical techniques using the classification criteria

In addition to statistics, scientists conducting research related to pharmaceutical practice and policy also employ techniques developed by the disciplines of biostatistics, econometrics, and psychometrics. Biostatistics and econometrics have applied standard statistical methods to their respective content fields (i.e., medical/health sciences and economics, respectively), but each also has advanced statistical theory and methods by addressing issues and concerns specific to its discipline (statisticians and econometricians do not always see eye to eye, e.g., Heckman 2005). Psychometrics, with its focus on educational and psychological measurement, has also applied and extended statistical methods, especially with respect to latent variable measurement and modeling. The following section introduces the tools developed by these disciplines in addition to the discipline of statistics, which are relevant to scientists who conduct pharmaceutical practice and policy-related research. Inevitably some techniques are excluded due to a limitation of space. For example, mostly missing from this chapter are a number of techniques falling under the general category of nonparametric or distribution-free statistics (Siegel and Castellan 1988; Higgins 2004).

Dependence techniques

Regression

There are many varieties of "regression analysis." Use of the unqualified term "regression" usually refers to linear regression with estimation performed using ordinary least squares (OLS) procedures. Thus, linear regression analysis is used for evaluating the relationship between one or more IVs (classically continuous, although in practice they can be discrete) and a single, continuous DV. As described elsewhere (e.g., see Muller and Fetterman [2002]), linear regression belongs to the general linear model (GLM) family of models.

Simple regression is used for the case of a single IV, whereas multiple regression refers to the case of more than one IV. When using multiple regression, researchers have the option of using computer-implemented, statistically based algorithms to aid in the addition and/or deletion of predictors from a regression model (often called variable selection procedures, such as stepwise regression). Although these techniques can be useful and informative, especially when used cautiously for purely predictive research or for exploratory research, they have limited application in theory-driven research, where order of entry of variables into a regression equation should be determined on theoretical grounds (researcher specification of a prior sequence for predictor variables is sometimes referred to as sequential or hierarchical regression). Multivariate regression implies the presence of multiple DVs, all continuous. Multivariate regression analysis can be viewed as path analysis or structural equation modeling with observed variables only (Raykov and Macoulides 2008), techniques to be addressed later. Such an analysis can also be performed using a technique called seemingly unrelated regression (SUR) (Beasley 2008).

There are a number of considerations when using linear regression, many of which revolve around the assumptions necessary for valid estimation and inference. The problems associated with assumption violations have led to the development of other techniques to address such violations. A full review of the GLM assumptions including violation detection, consequences, and remedies is beyond the scope of this chapter. However, there are excellent treatments of these topics, some more technical than others, in many different sources (e.g., Fox 1991; Muller and Fetterman 2002; Cohen et al. 2004; Kleinbaum et al. 2008).

Two other issues related to regression analysis deserve some consideration: (1) linear versus nonlinear models and (2) limited dependent variable models. One might hear the recommendation to use a nonlinear model. It is important to clarify what is meant by nonlinear. Several techniques falling under the category of generalized linear models (GzLMs, e.g., logistic, probit, and Poisson regression) are actually nonlinear regression models. These models involve the use of DVs that are discrete rather than continuous. In the context of linear regression, nonlinear may refer to the use of polynomial regression, where polynomial functions of predictors (e.g., X^2) are used in an attempt to estimate a nonlinear relationship between X and Y. The nonlinearity is in the predictors rather than in the parameters, so linear regression procedures can be used when polynomials are added to a regression equation (there are other considerations when conducting this type of analysis, see Cohen et al. [2004] and Kleinbaum et al. [2008]). Nonlinear transformations (e.g., logarithmic transformation) are also commonly used to linearize relationships, allowing for the use of linear regression procedures following transformation (although there may be some issues regarding interpretation). A nonlinear model may also refer to a model that is truly nonlinear in the parameters (referred to as intrinsically nonlinear such that the model cannot be linearized by transformation). These models require the use of nonlinear regression to estimate the coefficients. An example is certain types of exponential growth and decay models:

$$Y = \beta_0 + \beta_1[\exp(-\beta_2 X)] + E$$

Econometricians often use the term "limited dependent variables" to describe DVs that are limited in their range. Binary variables, count variables, and ordered responses are all examples of limited DVs. Techniques to address these situations are described later. Other mechanisms that restrict the range of DVs include censoring when truncation. Censoring occurs when the DV is set to some arbitrary value and the variable is beyond the censoring point (Baum 2006). In a cross-sectional survey on annual prescription drug purchases by individuals, those who made no purchases are censored at zero. Limits of measurement can also create censored data. With censored data, there are no observed values of the DV for those with values beyond the censoring point, but there is information about the predictor variables (censoring is also important in survival analysis). With truncation, neither the DV nor the IVs are observed for those with DV values that lie in the truncation region. Revisiting the prescription drug-purchasing scenario, data would be available from purchasers and nonpurchasers if collected in a cross-sectional survey. However, if data were collected from pharmacy sales records, they would include only purchasers and the data would be truncated.

Using OLS regression with such data, either truncated or censored, can create problematic estimates. Tobit regression was developed for censored data, whereas truncated regression models are available to address truncated data (Long 1997; Kennedy 2003; Baum 2006). It is also possible to use Heckman selection models with censored data (Kennedy 2003; Baum 2006).

Logistic regression and probit regression

Both logistic regression and probit regression are regression models used with discrete outcomes, classically a binary DV (i.e., dichotomous). Predictor variables can be continuous or discrete. They are both examples of GzLMs, of which the GLM is a special case. The difference between probit regression and logistic regression is technically what is called the link function; logistic regression uses the logit link function (natural log of the odds) whereas probit uses the probit link (related to the standard normal cumulative distribution function). The techniques often produce similar results and the decision about which to use may depend on the discipline in which the research was conducted and the desired interpretation of the parameter estimates. For example, probit is commonly used by toxicologists and to analyze dose–response data whereas logistic regression is popular among medical researchers and epidemiologists (it can be used to estimate odds ratios [an approximation to risk ratios] even from case–control studies) (Agresti 2007).

Both probit regression and logistic regression can be generalized to the cases of a response variable with three or more categories and when the response variable is ordered categories (i.e., ordinal response). Multinomial (polytomous) logistic regression and multinomial probit regression are used when a nominal DV has more than two categories. Ordinal logistic regression (also called ordered logit) and ordered probit regression are used when the categorical response categories are ordered.

Poisson regression

Poisson regression is another example of a GzLM and similar to logistic regression it is designed to analyze discrete data. It is used when the DV is a count variable, rather than a nominal or ordinal variable. Count data are quite common, especially in health-related research. The number of seizures in patients with epilepsy is an example of a count variable. Poisson regression is also referred to as a log-linear model. Overdispersion, having data exhibit greater variability than expected, is sometimes a problem when using Poisson regression. One alternative when this occurs is negative binomial regression, an extension of Poisson regression that better captures overdispersion (Agresti 2002).

Poisson regression is often used to model rate data, such as when an event occurrence is assessed over time or some other index of size, such as the number of cases of respiratory illnesses in a group of infants where each infant may be followed for a different length of time. Such models include something called an "offset" to account for different denominators in the rates (such as different observation periods in the previous example).

Zero-inflated and zero-truncated models are also available for both Poisson and negative binomial regression. Zero-inflated models can be used to address situations where there are excess zeroes in the data and zero-truncated models are used when the number of occurrences of an event is restricted to be positive. More information about these models is available in Long (1997).

Analysis of variance

In general, analysis of variance (ANOVA) involves assessing how one or more nominal IVs affect a continuous DV. The IVs in an ANOVA can be experimentally manipulated or intrinsic to the participants. ANOVA is strongly associated with the analysis of data from experimental designs. The typical use of ANOVA involves the comparison of several population means. When the focus is on a single IV, it is referred to as one-way ANOVA. The *t* test is a special case of a one-way ANOVA, a special case of multiple regression, which is a special case of the GLM. Almost any ANOVA model can be represented by a regression model using dummy variables.

A research design with two or more categorical IVs (factors), each studied at two or more levels, is called a factorial design. The situation of two IVs is called a two-way factorial design (higher-order designs with more factors are possible, but more difficult to interpret). Factorial designs allow for the assessment of whether the IVs are interacting with each other. An interaction occurs when the mean differences among the levels of one IV are not constant across the levels of the other IV. Or, more generally, an interaction is said to occur when the relationship between one IV and the DV depends on the value of the other IV.

Another important distinction when discussing ANOVA is whether the factor of interest is a between-participants factor or a within-participants factor. Factors in which each participant is measured at only one level of the IV are called between-participants factors. It is also possible that each participant is measured more than once (i.e., is exposed to multiple levels of the IV or is measured at multiple time points); such a factor is called a within-participants factor. Designs that have only within-participants factors are called within-participants designs (or repeated-measures designs). Repeated-measures ANOVA can be used to analyze such data, but other techniques are available.

Analysis of covariance

In general, analysis of covariance (ANCOVA) involves evaluating how one or more nominal IVs affect a continuous DV adjusted for the presence of control variable(s) called covariates (which are usually continuous). As with ANOVA models, ANCOVA models can be represented by a regression model by using dummy variables for the categorical IVs. ANCOVA is typically used to (1) reduce or eliminate systematic bias and (2) increase statistical power by removing variability from the error term.

It is possible to use a covariate in the cases of multiple IVs, called factorial ANCOVA, and it is also possible to use multiple covariates in an analysis. These are basic extensions of one-way ANCOVA with one covariate. ANCOVA is a frequently misused and abused technique, often applied with the hope that a quasi-experimental design can be made into an experimental design (e.g., see Miller and Chapman [2001] for a discussion of the issues). Myers and Well (2003), Keppel and Wickens (2004), and Maxwell and Delaney (2004) are excellent sources for a variety of ANOVA and ANCOVA applications.

Multivariate analysis of variance

Multivariate analysis of variance (MANOVA) is the multivariate extension of the univariate technique, ANOVA. In ANOVA, a single DV is tested for equality across the groups that comprise the IV. In the multivariate case, a variate (a linear combination of the multiple continuous DVs) is tested for equality. It is also possible to add additional IVs. With a single DV this is factorial ANOVA; with multiple DVs, this is factorial MANOVA. MANOVA can also be used with one or more covariates (this is called MANCOVA). Most multivariate statistics textbooks offer extensive treatments of MANOVA.

Discriminant function analysis

Discriminant function analysis (DA) is used when the DV is discrete and the IVs are continuous. The single DV can have two or more categories. DA can be used to describe group differences on the basis of the attributes (the IVs) of participants and to develop a rule that can be used to classify participants based on their IV scores (and potentially those with unknown group membership). DA and logistic regression can be used when the basic research question involves the prediction of group membership. They produce similar results when basic assumptions are met. Logistic regression has fewer assumptions and tends to be more robust to assumption violations than DA. Logistic regression can also accommodate discrete predictors whereas the basic assumptions of DA require continuous IVs (although it may be possible to relax this assumption when using DA for classification). Many researchers prefer logistic regression over DA because of its similarities to linear regression, including its statistical tests, incorporation of dummy variables, and the availability of many diagnostic procedures (Hair et al. 2006). However, when there are more than two groups, DA offers a certain richness with respect to interpretation.

There are many similarities between MANOVA and DA, although mathematically they are the same. DA is MANOVA turned around (Tabachnick and Fidell 2007); the IV in MANOVA (a grouping variable) becomes the DV in DA and the DVs in MANOVA (the continuous variables) become the predictor variables (i.e., the IVs) in DA. The difference between the two techniques lies in the direction of inference (and in a few other areas of application and interpretation, such as classification). If group differences are noted in MANOVA, the set of variables predicts group membership in DA.

Canonical correlation

Canonical correlation involves assessing the relationship between two sets of variables, a set of continuous DVs and a set of continuous IVs. When performing canonical correlation, linear combinations of the two sets of variables are developed that lead to the highest correlation between the two sets. For example, an educational researcher might assess the degree of relationship between a set of college performance variables and a set of precollege achievement variables. Both canonical correlation and multivariate multiple regression involve multiple continuous IVs and DVs and, although there are some differences in the objectives of the techniques, there are aspects of these two analyses that are mathematically equivalent (Lutz and Eckert 1994; Rencher 2002). Canonical correlation analysis is the most general of the multivariate techniques; multiple regression, MANOVA, discriminant analysis, and others are special cases of it (Tabachnick and Fidell 2007).

See Table 11.1 for a summary of the dependence techniques discussed thus far.

Conjoint analysis

The purpose of conjoint analysis is to determine the contribution of variables (and their levels) to individuals' judgments or choices. Although conjoint analysis has its roots in mathematical psychology (Luce and Tukey 1964) and econometrics (McFadden 1974), it has seen its greatest applications in marketing research, where, in general, it has been used to assess customer value for various attributes of products or services.

In determining the relative contribution of attributes on judgments or choices, conjoint analysis utilizes a decompositional approach, in which respondents provide their overall evaluations (ratings, rankings, or choices) of a set of profiles that are constructed by the researcher by combining levels of different attributes (IVs in the language of experimentation) (Green and Srinivasan 1978). Profiles supplied to respondents are constructed in such a

Table 11.1 Summary of several dependence techniques

Number and type of dependent variables	Number and type of independent variables		
	Discrete *X* (can be multiple)	Continuous *X* (can be multiple)	Discrete and continuous *X*s
Single discrete *Y*	Chi-squared tests	Logistic regression Probit regression Poisson regression Discriminant analysis	Logistic regression Probit regression Poisson regression
Single continuous *Y*	ANOVA (linear regression)	Linear regression	ANCOVA (linear regression)
Multiple continuous *Y*s	MANOVA	Canonical correlation Multivariate regression	MANCOVA

manner as to allow the importance of each attribute and each level of each attribute to be determined from the respondents' overall responses (Hair et al. 2006). Conjoint analysis can also be used to predict the demand for a product or service with given attributes (Hair et al. 2006).

There are different conjoint methodologies, including the traditional methodology which relies on ratings and rankings, choice-based conjoint analysis which is used for discrete choice modeling, and adaptive conjoint analysis which is used when there are many different attributes under study. Use of conjoint analysis in healthcare is quite prevalent as researchers have attempted to understand how patients and others value aspects of not only healthcare services and products but also their health. Overviews of conjoint analysis and discrete choice experiments and their applications to healthcare can be found in Ryan and Farrar (2000) and Ryan and Gerard (2003).

Path analysis

Path analysis (PA) is an extension of multiple regression in which a system of simultaneous equations among observed (or manifest) variables is modeled. Thus, there may be multiple DVs, IVs, and other variables known as intervening variables that transmit some or all of the effects of an antecedent variable to a subsequent variable (this intervening variable effect goes by names such as mediation effect, indirect effect, and intermediate outcome effect). In most applications of PA, the effects are unidirectional; however, it is possible to incorporate feedback loops (reciprocal causation). These models are examples of nonrecursive models, and the analysis and the interpretation are sufficiently more difficult than recursive models (see Kline [2005, 2006] for a discussion).

Researchers using PA are guided by theory when constructing the model relating the variables of interest. PA then allows for an estimation of presumed causal relationships among variables. This has led some to refer to it (as well as the more general technique of structural equation modeling) as causal modeling. This has generated considerable discussion in the literature, especially as PA (and structural equation modeling or SEM) are usually (although do not have to be) applied to data collected through observational studies. It is possible to assess how well the proposed model fits the data. Poor fitting models can be rejected, but, if the model and the data are consistent, all that can be said is that there is evidence to support the theoretical model proposed by the researcher; it does not prove that the proposed model is correct.

Structural equation modeling

PA is really a special case of SEM. Indeed, many of the techniques discussed so far, including multiple regression and ANOVA, are special cases of SEM. SEM is one of the most widely used statistical methodologies in the social sciences. The general topic of SEM is so vast that it is difficult to structure a paragraph or two discussing its applications. Numerous books and journal articles addressing general SEM use, as well as specific issues, have been published. Traditionally, SEM can be thought of as a "melding of factor analysis and path analysis into one comprehensive statistical methodology" (Kaplan 2009). Thus, with SEM it is possible to estimate path models (called structural models) that involve latent variables. Such analyses have the advantage of potentially correcting for the adverse effects of measurement error (i.e., relationships among latent variables will be disattenuated for measurement error). In addition to fitting structural models with latent variables, traditional coverage in SEM books and courses typically includes PA (structural models with observed variables only), confirmatory factor analysis (used to evaluate measurement models, which link latent variables to manifest variables), and multiple-group analyses (which can be used to assess measurement invariance, interaction effects, and differences in latent variable means between known groups).

A number of more advanced structural equation models can be utilized with longitudinal data. Admitting models for categorical latent variables into the SEM framework adds a number of interesting applications. Finally, working with categorical manifest variables has also extended the tools available to SEM users.

Survival analysis

Survival analysis is actually a group of "statistical methods for studying the occurrence and timing of events" (Allison 1995). In medical settings, the event of interest is often death, which is why this technique is commonly referred to as survival analysis. More generally, the class of techniques is referred to as time-to-event analysis and goes by different names in different academic disciplines such as duration modeling (economics), event history analysis (sociology), and reliability or failure time analysis (engineering) (Allison 1995; Kennedy 2003). Although the event is typically labeled as a failure, survival analysis can be used to study time to event for just about any outcome, positive or negative (e.g., graduation, child birth). A key feature with time-to-event data is that events will not necessarily have occurred for the entire sample, meaning that the full survival time is not known for some individuals, a problem known as censoring. Survival analysis techniques are designed to address censoring.

Survival analysis techniques can be classified as continuous-time methods and discrete-time methods. In general, discrete-time methods are used when the events can occur only at specific moments (e.g., researchers know only the year in which an event occurred, such as in a study of turnover among teachers, as most teachers who quit leave at the end of the school year). Discrete-time methods often rely on techniques for binary DVs, such as logistic regression and probit regression (Allison 1995). There are a variety of continuous-time methods. The Kaplan–Meier (KM) method is a nonparametric procedure for estimating what is called a survivor function. Plots of KM survival curves for treatment and control groups are very common in the medical literature. Parametric regression models for censored survival data include accelerated failure time (AFT) models such as the exponential and the Weibull models, and allow researchers to estimate parameters describing the relationship between predictors and time to the event. Finally, semi-parametric models, generally known as Cox's regression models (proportional hazards models), can do many of the same things as the parametric procedures (and more) but without having to choose a particular probability distribution to represent survival times.

Survival analysis methods continue to expand because this is an active area of statistical research. Inclusion of time-dependent covariates, the modeling of competing risks, evaluating multiple (or repeated) events, and evaluating the predictive ability of survival analysis models are topics that continue to receive attention from statisticians and applied researchers. There are many resources available to learn more about survival analysis, including Singer and Willett (2003) and Hosmer et al. (2008).

Time-series analysis

Time-series data represent data for a single entity (could be a person, a firm, a county, a country, etc.) collected repeatedly over many periods of time. This is in contrast to data collected from multiple entities at multiple time points (referred to as panel data and other names). Both represent situations of correlated data; the second is discussed later. Time-series analysis has roots in both the statistics literature and the economics literature (Kennedy 2003), and can be used to create forecasts (i.e., predicting future values), test for seasonality and other trends, examine the effects of an intervention (such as a new policy, natural event), and explore the relationship of two (or more) variables measured over time. The techniques used in time-series analysis are many and varied, and include autoregression, distributed lag models, autoregressive distributed lag models, autoregressive integrated moving average (ARIMA) models, and vector autoregression (VAR), which extends the univariate autoregression model to the case of multiple variables. Most econometric textbooks cover time-series analysis and forecasting. A review of some of these techniques written for social scientists is provided by Tabachnick and Fidell (2007).

Interdependence techniques

Factor analysis

Factor analysis attempts to define the underlying structure among a set of variables. It is commonly used in the development and refinement of measurement scales. Exploratory factor analysis (EFA) is "an inductivist method designed to discover an optimal set of factors, their number to be determined in the analysis, that accounts for the covariation among the items" (Skrondal and Rabe-Hesketh 2004). Any indicator (i.e., variable, item) may be associated with any factor. There is no prior theory about the underlying structure.

With confirmatory factor analysis (CFA), theory is used to guide the construction of a measurement model (i.e., specifying the number of factors, which measured variables load on which factor). SEM is usually used to conduct CFA by assessing the fit of proposed measurement models. With factor analysis, not only are the latent variables assumed to be continuous, but so are the manifest (or indicator) variables. The use of categorical (or discrete) indicator variables has been the subject of much debate and research in the SEM and measurement literatures. CFA with the categorical indicators is referred to as item response theory (IRT) analysis (also called latent trait analysis) (Takane and de Leeuw 1987; Kamata and Bauer 2008). The latent variable in an IRT analysis remains continuous. IRT (also called latent trait theory) has a rich history (Embretson and Reise 2000). Although there is some disagreement, Rasch models can be considered a subset of IRT models.

Revisiting EFA, there is technically a difference between factor analysis and another technique commonly discussed under the factor analysis umbrella – principal components analysis (PCA). With PCA, all of the variance in the observed variables is analyzed, but with factor analysis only common or shared variance is analyzed. In addition, factors are thought to "cause" observed variables, whereas components are simply aggregates, or linear combinations, of correlated observed variables. The variables are the "cause" of the component. Some view PCA as a special kind of factor analysis, whereas others maintain that they are very different (Hair et al. 2006).

Resources for more information about EFA and sometimes CFA include most multivariate statistics textbooks. SEM textbooks cover the basics of CFA and sometimes more advanced applications; Brown (2006) is another useful reference.

Cluster analysis

Generally, cluster analysis is a technique used for combining observations (these observations can be individuals, firms, products, countries, etc.) into homogeneous groups or clusters based on a set of variables. Group membership is not known beforehand and the purpose of the analysis is to develop groups on the basis of similarities among the observations in the sample on a set of characteristics. Although it is a very useful technique with many applications, as Hair et al. (2006) note: "cluster analysis is descriptive, atheoretical, and noninferential ... it has no statistical basis upon which to draw inferences from a sample to a population, and many contend that it is only an exploratory technique." There are a number of different approaches to forming clusters generally classified as either hierarchical or nonhierarchical (also called relocation or partitioning methods). An example of a hierarchical algorithm is Ward's method. K-means clustering is a commonly used nonhierarchical procedure. For additional coverage of the many issues involved in traditional cluster analysis see Hair et al. (2006) and Lattin et al. (2003).

Latent class analysis

As Vermunt and Magidson (2004) observe: "the basic idea underlying latent class (LC) analysis is a very simple one: some of the parameters of a postulated statistical model differ across unobserved subgroups." In the statistics literature (as opposed to the social science literature), LC models are referred to as finite mixture models (Muthén 2002; Vermunt and Magidson 2004). In its traditional form, LC analysis (LCA) was used for creating typologies (or discrete unknown classes or clusters) based on nominal (discrete) variables (Vermunt and Magidson 2004). Thus the classes can be thought of as the categories of a categorical latent variable (Skrondal and Rabe-Hesketh 2004). As opposed to factor analysis and latent trait analysis, the latent variable in LCA is discrete. It is possible to add covariates to the models, predicting latent class membership with a set of predictor variables of interest typically using multinomial logistic regression (Muthén 2002).

LCA has also been extended to the case of continuous indictors (it is possible to have mixed variable types – continuous and discrete). This is generally referred to as latent profile analysis and by a variety of other names such as mixture-model clustering, model-based clustering, and LC clustering (Vermunt and Magidson 2004). The primary difference between this type of cluster analysis and traditional cluster analysis described previously is that a probability-based distance measure is used rather than the ad hoc distance measures employed by traditional cluster algorithms. Cases in the same latent class "are similar to each other because their responses are generated by the same probability distribution" (Magidson and Vermunt 2004). LCA-based clustering techniques have been compared with traditional clustering techniques, such as K-means clustering (Magidson and Vermunt 2002).

Table 11.2 provides a fourfold classification table of several latent variable models discussed thus far (Bartholomew and Knott 1999).

LCA can also be "combined" with other dependence and interdependence techniques. For example, regression mixture analysis (also called latent class regression analysis) allows for the estimation of separate regression equations for each latent class. Regression coefficients are allowed to vary across the latent classes (in essence, the LC variable serves as a moderator, involving the presence of an interaction – see Ding [2006] for an example). This basic idea can be extended to mixture modeling with other techniques (e.g., structural equation mixture modeling, path analysis mixture models, and factor mixture analysis). A number of other models can be utilized with longitudinal data, some of which will be mentioned when discussing the modeling of correlated data. Most of these models (if not all) can be fit with the M*plus* program (Muthén and Muthén, Los Angeles, CA), a very flexible program for addressing many different kinds of latent variable models. Latent GOLD (Statistical Innovations, Belmont, MA) can also fit many of these models.

Table 11.2 Summary of latent variable models – interdependence techniques

Manifest variables	Latent variable(s)	
	Continuous	Discrete
Continuous	Factor analysis	Latent profile analysis
Discrete	Latent trait analysis	Latent class analysis

Multidimensional scaling

Multidimensional scaling (MDS) is really a set of methods that attempts to reveal dimensions underlying similarities or distances among objects (in marketing these objects are often products or brands, but they can be just about anything, such as political candidates or issues, cities or cultures, journals, or even individuals themselves). The output is often presented in something called a perceptual map, which provides coordinates (interval-scaled variables) of each object in multidimensional space (the usual presentation is a two-dimensional grid). Indeed, MDS is sometimes referred to as perceptual mapping, although it is probably best thought of as a type of perceptual mapping. In its typical application, MDS is used when the dimensions are not known beforehand but rather are derived from global judgments provided by individuals (a decompositional or attribute-free approach). These may be judgments of similarity or preference (the choice depends on the nature of the research question). When objective measures on a defined set of attributes are available for a set of objects, cluster analysis is the preferred technique (although the output is cluster membership, a nominal variable, rather than location in multidimensional space, an interval variable). With MDS, it is possible to produce perceptual maps at the individual level (disaggregate analysis) or a single combined map for all respondents (aggregate analysis). See Lattin et al. (2003) and Hair et al. (2006) for more discussion of MDS.

Modeling correlated data

The last classification criterion, the presence of correlated data, deserves special consideration. Correlated data are quite common in almost all types of research. Such data typically occur when units are nested within clusters. Examples from different disciplines include: students within classes (educational research), employees within organizations (business research), patients within hospitals (medical research), and animals within litters (biology). The data can come from multiple levels (e.g., variables at the employee level and organizational level) and the data structure is said to be hierarchical (note that there can be more than two levels). This is why two common names associated with the analysis of such data are multilevel analysis and hierarchical data modeling (not to be confused with order of entry determination in multiple regression, sometimes called hierarchical regression). These models are also called random coefficient models as the coefficients (i.e., intercepts, slopes) are allowed to vary across higher-level units (Tabachnick and Fidell 2007).

Multilevel analysis allows researchers to examine the effects of variables at multiple levels and even cross-level interactions. In addition, units within clusters tend to be more like each other than units from two different clusters. This is what it means to have correlated data and it leads to a violation of the independence assumption that applies to most of the techniques discussed thus far. Analyses of multilevel data need to take this correlation into account. Ignoring the hierarchical data structure can lead to both interpretational and statistical errors (Tabachnick and Fidell 2007).

Longitudinal data (also called repeated-measures, panel, or cross-sectional time-series data) are another example of correlated data and can be viewed in terms of a hierarchy with multiple levels, i.e., multiple observations nested within individuals. The techniques of multilevel modeling can be used to analyze these types of data. This approach is more flexible and offers several advantages over more traditional approaches to the analysis of repeated measures data such as repeated measures ANOVA (univariate approach) and the multivariate analysis of variance approach (sometimes called profile analysis). It is referred to by a variety of names, including individual growth modeling and the multilevel model for change. Many multilevel analysis books have chapters on longitudinal data analysis (e.g., Snijders and Bosker 1999; Hox 2002) and Singer and Willett (2003) provide an excellent treatment of the multilevel model for change.

When the DV of interest is continuous, these models are referred to as linear mixed models, hierarchical linear models, or multilevel linear models. Just as the general linear model for continuous DVs can be extended to handle discrete DVs, so the linear mixed model can be extended to situations involving a discrete DV. These models are called generalized linear mixed models (GLMMs) or hierarchical generalized linear models, of which the linear mixed model is a special case. The method of generalized estimating equations (GEEs) is also used in this context, especially for longitudinal data. There is some debate over which method to use especially as, in the case of discrete DVs, GEE and GLMM estimates and their interpretations are different (for a discussion of population-averaged effects versus subject-specific effects, see Agresti [2007]). Multilevel analysis can be generalized to many different techniques, such as path analysis, factor analysis, survival analysis, and SEM (Muthén and Asparouhov 2010).

There are stand-alone software packages for conducting multilevel analysis (e.g., HLM, Scientific Software International, Lincolnwood, IL), as well as routines in the major comprehensive statistical software packages. In addition to the approaches discussed above, fixed effects regression methods are commonly used in other disciplines, especially for panel data (Allison 2005, 2009). In the healthcare literature, the fixed-effects approach commonly utilized is conditional logistic regression especially for matched case–control studies.

The consideration of latent variables in the longitudinal setting adds another element and allows researchers to address additional questions. There are several examples of such models. Latent growth models are an alternate approach to multilevel models for measuring change. Indeed, under certain conditions, the two approaches are identical (Curran 2003; Hox and Stoel 2005). Latent growth modeling offers some additional flexibility in some areas (Singer and Willett 2003). Latent transition analysis, latent class growth analysis, and growth mixture analysis are possible with categorical latent variables. Latent transition analysis examines how individuals' latent class memberships change over time, i.e., individuals can "transition" between latent classes (e.g., see Kaplan 2008). Latent class growth analysis and growth mixture analysis allow for the growth curves to vary across unobserved subpopulations (i.e., latent classes) (Muthén 2004).

Studies using complex sample designs, commonly used when conducting large sample surveys (see Chapter 13), can also create correlated data. Design features might include cluster sampling, stratification, and disproportionate sampling schemes (e.g., oversampling of some groups). Many authors distinguish between design-based and model-based perspectives in the analysis of complex survey data (Lee and Forthofer 2006). Design-based approaches are quite common and are implemented in programs such as SUDAAN (Research Triangle Institute, Research Triangle Park, NC). These programs also incorporate weights to account for differential representation in the sample due to unequal probabilities of selection. Such approaches can be used for many types of analyses, including descriptive analysis, regression analysis, and even SEM (e.g., see Stapleton 2006, 2008). Multilevel analysis can also be used to take into account clustering and allows for an examination of variables from different levels simultaneously; design-based analyses are usually not interested in the latter. The use of multilevel modeling to analyze survey data generated with complex sampling is an example of a model-based approach.

As with SEM, the literature surrounding the analysis of correlated data is so vast that it is almost impossible to succinctly summarize all of it. This is still an area of active research in the statistics community and it is entirely possible that new approaches will become available with time.

Selecting an appropriate statistical technique

Which tool one chooses for an analysis depends on the classification criteria discussed earlier. As pointed out several times, many of the techniques discussed are related; some are special cases of more general techniques. It is important to try to understand such relationships because it helps to interpret analyses, but also to understand a wider variety of techniques. For a large umbrella framework of statistical techniques see Skrondal and Rabe-Hesketh (2004). Just about all the cases in the above discussion, with some exceptions, are special cases of this framework.

Researchers may find that there are often multiple approaches to address the same general research question (e.g., DA, logistic regression, and probit regression). In these situations, it is critical to know the research area (as well as the target journal) and understand the techniques that are commonly used in specific disciplines in which the work is primarily based. One other important consideration is that researchers should design the analysis plan to answer the research question; investigators should not develop the research questions to match the type of analysis to be conducted.

Summary and conclusions

This chapter introduces the reader to some statistical terminology and to many different approaches to analyzing data. It is impossible for a single chapter to cover the many issues that receive attention in both the statistical literature and the literature of specific content areas; thus, many current topics in statistics have not been addressed. Therefore, a number of references, both primary literature and textbooks, have been provided for interested readers to deepen their understanding of particular areas.

As mentioned in the chapter introduction, scientists have access to many different sets of tools to help make sense of the world and the statistics toolbox plays a very important role in the generation and testing of new knowledge. It is imperative that, before using these tools, more study be undertaken. Some applications of the techniques described in this chapter may lead to results that are difficult to interpret. In addition, sometimes model fitting can be very challenging. To extend the toolbox analogy, this chapter described the contents of the toolbox – the reader might still have to study the owners' manual before employing these tools! Hopefully, the introduction and overview provided in this chapter will stimulate the reader to learn more about how many of these techniques are related to each other, how to appropriately use these tools, and most importantly how these methods can help to address substantive research questions.

Review questions/topics

1 What are the two major functions served by statistics?
2 Describe the various schemes used to classify variables. Why are such classification systems important to understanding the applications of statistical methods?
3 What is the difference among linear regression, logistic regression, Poisson regression, and Cox's regression?
4 A reviewer asks you to consider the use of a nonlinear model. What might she mean?
5 What is the difference among ANOVA, ANCOVA, and MANOVA?
6 What does it mean that path analysis is a special case of structural equation modeling?

> **7** What is the difference between exploratory and confirmatory factor analysis?
>
> **8** How do traditional cluster analysis and latent class analysis differ?
>
> **9** Under which situations would a researcher need to consider the use of techniques that handle correlated data?
>
> **10** What are some latent variable techniques that can be used with longitudinal data?

References

Agresti A (2002). *Categorical Data Analysis*, 2nd edn. Hoboken, NJ: Wiley.

Agresti A (2007). *An Introduction to Categorical Data Analysis*, 2nd edn. Hoboken, NJ: Wiley.

Allison PD (1995). *Survival Analysis using SAS: A practical guide*. Cary, NC: SAS Institute.

Allison PD (2005). *Fixed Effects Regression Methods for Longitudinal Data Using SAS*. Cary, NC: SAS Institute.

Allison PD (2009). *Fixed Effects Regression Models*. Thousand Oaks, CA: SAGE Publications.

Bartholomew DJ, Knott M (1999). *Latent Variable Models and Factor Analysis*. London: Arnold.

Baum GF (2006). *An Introduction to Modern Econometrics Using Stata*. College Station, TX: StataCorp LP.

Beasley TM (2008). Seemingly unrelated regression (SUR) models as a solution to path analytic models with correlated errors. *Multiple Linear Regression Viewpoints* 34: 1–7.

Bollen KA (2002). Latent variables in psychology and the social sciences. *Annu Rev Psychol* 53: 605–34.

Brown TA (2006). *Confirmatory Factor Analysis for Applied Research*. New York: The Guilford Press.

Cohen J, Cohen P, West SG, Aiken LS (2004). *Applied Multiple Regression/Correlation Analysis for the Behavioral Sciences*, 3rd edn. Mahwah, NJ: Lawrence Erlbaum Associates.

Curran PJ (2003). Have multilevel models been structural equation models all along? *Multivariate Behav Res* 38: 529–69.

Ding CS (2006). Using regression mixture analysis in educational research. *Practical Assessment, Research Evaluation* 11(11). Available at: http://pareonline.net/getvn.asp?v=11&n=11http://pareonline.net/getvn.asp?v=11&n=11.

Embretson SE, Reise SP (2000). *Item Response Theory for Psychologists*. Mahwah, NJ: Lawrence Erlbaum Associates.

Fox J. (1991). *Regression Diagnostics*. Newbury Park, CA: SAGE Publications.

Green PE, Srinivasan V (1978). Conjoint analysis in consumer research: Issues and outlook. *J Consumer Res* 5: 103–23.

Grootendorst P (2007). A review of instrumental variables estimation of treatment effects in the applied health sciences. *Health Serv Outcomes Res Methodol* 7: 159–79.

Hair JF, Black WC, Babin BJ, *et al.* (2006). *Multivariate Data Analysis*, 6th edn. Upper Saddle River, NJ: Prentice Hall.

Harlow LL, Mulaik SA, Steiger JH, eds (1997). *What if There were No Significance Tests?* Mahwah, NJ: Lawrence Erlbaum Associates.

Heckman JJ (2005). Rejoinder: Response to Sobel. *Sociol Methodol* 35: 135–50.

Higgins JJ (2004). *Introduction to Modern Nonparametric Statistics*. Pacific Grove, CA: Brooks/Cole-Thomson Learning.

Hosmer DW, Lemeshow S, May S (2008). *Applied Survival Analysis: Regression modeling of time-to-event data*, 2nd edn. Hoboken, NJ: Wiley.

Hox J (2002). *Multilevel Analysis: Techniques and applications*. Mahwah, NJ: Lawrence Erlbaum Associates.

Hox J, Stoel RD (2005). Multilevel and SEM approaches to growth curve modeling. In: Everitt BS, Howell DC (eds), *Encyclopedia of Statistics in Behavioral Science*. Chichester: Wiley, 1296–305.

Kamata A, Bauer DJ (2008). A note on the relation between factor analytic and item response theory models. *Structural Equation Modeling* 15: 136–53.

Kaplan D (2008). An overview of Markov chain methods for the study of stage-sequential developmental processes. *Dev Psychol* 44: 457–67.

Kaplan D (2009). *Structural Equation Modeling: Foundations and Extensions*, 2nd edn. Thousand Oaks, CA: SAGE Publications.

Katz MH (2003). Multivariable analysis: A primer for readers of medical research. *Ann Intern Med* 138: 644–50.

Kennedy P (2003). *A Guide to Econometrics*, 5th edn. Cambridge, MA: The MIT Press.

Keppel G, Wickens TD (2004). *Design and Analysis: A researcher's handbook*, 4th edn. Upper Saddle River, NJ: Prentice Hall.

Kleinbaum DG, Kupper LL, Nizam A, Muller KE (2008). *Applied Regression Analysis and Other Multivariable Methods*, 4th edn. Belmont, CA: Thomson Brooks/Cole.

Kline RB (2005). *Principles and Practice of Structural Equation Modeling*, 2nd edn. New York: The Guilford Press.

Kline RB (2006). Reverse arrow dynamics: Formative measurement and feedback loops. In: Hancock GR, Mueller RO (eds), *Structural Equation Modeling: A second course*. Greenwich, CT: Information Age Publishing, 43–68.

Lattin JM, Carroll JD, Green PE (2003). *Analyzing Multivariate Data*. Pacific Grove, CA: Thomson Brooks/Cole.

Lee ES, Forthofer RN (2006). *Analyzing Complex Survey Data*, 2nd edn. Thousand Oaks, CA: SAGE Publications.

Long JS (1997). *Regression Models for Categorical and Limited Dependent Variables*. Thousand Oaks, CA: Sage Publications.

Luce RD, Tukey JW (1964). Simultaneous conjoint measurement: A new type of fundamental measurement. *J Math Psychol* 1: 1–27.

Lutz JG, Eckert TL (1994). The relationship between canonical correlation analysis and multivariate multiple regression. *Educ Psychol Meas* 54: 666–75.

McFadden D (1974). Conditional logit analysis of qualitative choice behavior. In: Zarembka P (ed.), *Frontiers in Econometrics*. New York: Academic Press, 105–42.

Magidson J, Vermunt JK (2002). Latent class models for clustering: A comparison with K-means. *Can J Marketing Res* 20: 37–44.

Magidson J, Vermunt JK (2004). Latent class models. In: Kaplan D (ed.). *The Sage Handbook of Quantitative Methodology for the Social Sciences*. Thousand Oaks: Sage Publications, 175–98.

Maxwell SE, Delaney HD (2004). *Designing Experiments and Analyzing Data: A model comparison perspective*, 2nd edn. Mahwah, NJ: Lawrence Erlbaum Associates.

Miller GA, Chapman JP (2001). Misunderstanding analysis of covariance. *J Abnorm Psychol* 110: 40–8.

Muller KE, Fetterman BA (2002). *Regression and ANOVA: An integrated approach using SAS software*. Cary, NC: SAS Institute.

Muthén BO (2002). Beyond SEM: General trait latent variable modeling. *Behaviormetrika* 29: 81–117.

Muthén BO (2004). Latent variable analysis: Growth mixture modeling and related techniques for longitudinal data. In: Kaplan D (ed.), *The Sage Handbook of Quantitative Methodology for the Social Sciences*. Thousand Oaks, CA: Sage Publications, 345–68.

Muthén B, Asparouhov T (2010). Beyond multilevel regression modeling: Multilevel analysis in a general latent variable framework. In: Hox J, Roberts JK (eds), *The Handbook of Advanced Multilevel Analysis*. Basingsoke: Taylor & Francis; in press. Available at: www.statmodel.com/download/multilevelVersion2.pdf.

Myers JL, Well AD (2003). *Research Design and Statistical Analysis*, 2nd edn. Mahwah, NJ: Lawrence Erlbaum Associates.

Nickerson RS (2000). Null hypothesis significance testing: A review of an old and continuing controversy. *Psychol Methods* 5: 241–301.

Norman GR, Streiner DL (2008). *Biostatistics: The bare essentials*, 3rd edn. Shelton, CT: People's Medical Publishing House.

Normand ST (2008). Some old and some new statistical tools for outcomes research. *Circulation* 118: 872–84.

Raykov T, Marcoulides GA (2008). *An Introduction to Applied Multivariate Analysis*. New York: Routledge.

Rencher AC (2002). *Methods of Multivariate Analysis*, 2nd edn. Hoboken, NJ: Wiley.

Rubin DB (1997). Estimating causal effects from large data sets using propensity scores. *Ann Intern Med* 127: 757–63.

Ryan M, Farrar S (2000). Using conjoint analysis to elicit preferences for health care. *BMJ* 320: 1530–3.

Ryan M, Gerard K (2003). Using discrete choice experiments to value health care programmes: Current practice and future research reflections. *Appl Health Econ Health Policy* 2: 55–64.

Schneeweiss S (2007). Developments in post-marketing comparative effectiveness research. *Clin Pharmacol Ther* 82: 143–56.

Siegel S, Castellan NJ (1988). *Nonparametric Statistics for the Behavioral Sciences*, 2nd edn. New York: McGraw-Hill.

Singer JD, Willett JB (2003). *Applied Longitudinal Data Analysis: Modeling change and event occurrence*. Oxford: Oxford University Press.

Skrondal A, Rabe-Hesketh S (2004). *Generalized Latent Variable Modeling: Multilevel, longitudinal, and structural equation models*. Boca Raton, FL: Chapman & Hall/CRC.

Snijders T, Bosker R (1999). *Multilevel Analysis: An introduction to basic and advanced multilevel modeling*. London: Sage Publications.

Stapleton LM (2006). An assessment of practical solutions for structural equation modeling with complex sample data. *Structural Equation Modeling* 13: 28–58.

Stapleton LM (2008). Variance estimation using replication methods in structural equation modeling with complex sample data. *Structural Equation Modeling* 15: 183–210.

Tabachnick BG, Fidell LS (2007). *Using Multivariate Statistics*, 5th edn. Boston, MA: Allyn & Bacon.

Takane Y, de Leeuw J (1987). On the relationship between item response theory and factor analysis of discretized variables. *Psychometrika* 52: 393–408.

Vermunt JK, Magidson J (2004). Latent class analysis. In: Lewis-Beck M, Bryman AE, Liao TF (eds), *The Sage Encyclopedia of Social Science Research Methods*. Thousand Oaks, CA: SAGE Publications, 549–53.

Wunsch H, Linde-Zwirble WT, Angus DC (2006). Methods to adjust for bias and confounding in critical care health services research involving observational data. *J Crit Care* 21: 1–7.

Online resources

The Academic Technology Services. University of California, Los Angeles. Available at: www.ats.ucla.edu/stat.

The Division of Statistics & Scientific Computation. University of Texas at Austin. Available at: http://ssc.utexas.edu/consulting/answers.

There are many excellent online resources regarding statistical analysis. All of the major software vendors (i.e., SPSS, SAS, STATA, M*plus*, etc.) have excellent information available on their websites.

12

Secondary data analysis: administrative data

Bradley C Martin

Chapter objectives

- To define administrative data
- To describe Medicaid and Medicare data
- To describe analytical files and data structure in claims data
- To discuss common medical coding conventions
- To examine use of administrative data for research

Defining administrative data

Administrative health data are those collected, recorded, or transmitted primarily for payment or management purposes, and are not specifically designed for research uses. The most common health administrative data sources used by researchers are the records of financial transactions among health providers, payers, and patients, and are routinely referred to as administrative claims data. These data are used primarily for payment purposes where providers are reimbursed in a fee-for-service system and to record services and products delivered to a patient, including many financial details associated with the service or product, as well as general characteristics of the service or product being delivered. Examples include administrative claims data, such as Medicare and Medicaid from governmental agencies, and commercial insurance data, such as Humana and Aetna. The primary focus of this chapter is on Medicare and Medicaid administrative data and Chapter 13 discusses commercial data sources for research.

With more health providers adopting the use of electronic medical records (EMRs), essentially recording information that would appear in a patient's chart, these data will represent a growing source of administrative data that can be used alone or in conjunction with administrative claims data. Other sources of administrative data are summary information collected by a group of providers such as hospitals or emergency rooms (ERs), which make available hospital discharge abstracts and ER encounters. Additional administrative data that may have uses in the healthcare setting might include census data as well as aggregate data on health spending. There is also an array of surveys collected specifically for research purposes and the Medical Expenditure Panel Survey (MEPS) or several surveys of providers and patients conducted by the National Center for Health Statistics (NCHS). These surveys are discussed in Chapter 14.

This chapter focuses on research application of Medicare and Medicaid data. Specifically, the analytical files, data structure, and medical coding conventions are discussed to provide an analytic framework for pharmaceutical practice and policy research. Although the research approaches are presented in the framework of Medicare and Medicaid data, many of these approaches are equally applicable to any administrative claims data.

Uses of administrative claims data

As administrative data are characterizations of medical encounters, services delivered, or products used in the past, they are by definition retrospective and observational in nature. In the research design sense, retrospective refers to past occurrences, and observational to where the researcher is simply observing what naturally unfolds and cannot control or manipulate which patients are exposed to a particular stimulus, such as a specific drug or procedure. There is a wide array of uses of administrative health claims data that range from policy analyses, comparative effectiveness research, pharmacoepidemiology and drug safety, to forecasting future health spending, cost of illness, and adherence studies, to name but a few. Research using these data cannot be used to establish the efficacy of a drug, product, or procedure. Efficacy studies require randomized controlled trials (RCTs) and studies based exclusively on administrative data cannot be used to acquire a labeled indication with the US Food and Drug Administration (FDA).

One of the ongoing debates in the health field is whether studies based on administrative data can be used to inform specific clinical decisions regarding the use of drugs or the effectiveness of procedures. Due to the limitations inherent in observational retrospective studies, some clinicians view studies based on administrative data as hypothesis-generating studies that need to be confirmed with RCT data before clinicians can alter their prescribing practices. Given the lack of relatively few head-to-head trials within a therapeutic class, others feel that there may be sufficient levels of evidence provided by these studies. Recently there has been growing interest in using these data for comparative effectiveness research to inform clinical prescribing and the American Recovery and Reinvestment Act of 2009 provided US$1.1 billion toward that goal (US Department of Health and Human Services 2009).

In 2009, the Institute of Medicine convened a panel to identify US priorities for comparative effectiveness research and developed a list of the top 100 research priorities, of which 28 could be addressed using retrospective database resources.

Medicare and Medicaid data sources

Medicaid claims data

Medicaid, or title XIX of the Social Security Act, was enacted in 1965 to cover the healthcare services of indigent and disabled individuals (Centers for Medicare and Medicaid Services or CMS no date, a). Each state in the USA operates and manages its own Medicaid program under federal guidelines and oversight. As a result, there are no two identical Medicaid programs and there is a fair amount of heterogeneity between the state programs with regard to the recipients enrolled and the benefits available, but it is fair to say that all Medicaid programs provide a reasonably comprehensive set of health benefits to the states' indigent and disabled individuals. Medicaid recipients tend to be infants and children, young women, poor elderly people, or disabled individuals with high health needs because of the eligibility rules for Medicaid programs. As Medicaid systems generally offer a comprehensive set of health benefits, these data are a rich resource for studying a wide range of research questions for these populations, but are not well suited for studying disease or drug use patterns in nondisabled middle-aged adults.

Medicaid recipients

Before starting a research investigation based on Medicaid claims data, one must become familiar with the individuals who qualify for Medicaid services and the benefits afforded to the various beneficiaries. There are three broad categories of individuals covered by Medicaid: the categorically needy, the medically needy, and the optional eligibles (CMS 2005). Categorically needy recipients are people eligible for Medicaid who have incomes below state thresholds and meet other requirements based on age, pregnancy, or disability. Most categorically needy recipients will receive "cash assistance" although some may not. Medically needy recipients "spend down" to Medicaid eligibility by incurring medical and/or remedial care expenses to offset their excess income, thereby reducing it to a level below the maximum allowed by that state's Medicaid plan. Optional eligibles are people who become eligible for Medicaid services who would not qualify for any of the above categories and may be included in Medicaid waiver programs expanding eligibility to more targeted recipients. Federal guidelines set out provisions that a state Medicaid program must cover, which include services to the following individuals: limited income families with children, temporary assistance for needy families (TANF), supplemental security income (SSI) recipients (aged, blind, and disabled individuals), children under age 19, and pregnant women whose family income is at or below 133% of the federal poverty level (FPL), recipients of adoption assistance, and foster care infants born to Medicaid-eligible pregnant women (CMS no date, a).

Although not directly a Medicaid program, the State Children's Health Insurance Program (SCHIP) is administered by the Medicaid program in most states, and the claims data for these children are sometimes included with the Medicaid claims data. The SCHIP program expanded health insurance to children whose families earn too much money to be eligible for Medicaid, but not enough to purchase private insurance, and was enacted in 1997 as part of the Balance Budget Act. States determine the design of its program, eligibility groups, benefit packages, payment levels for coverage, and administrative and operating procedures. The SCHIP provides coverage to children residing in households that have incomes between 150 and 350 percent of the FPL based on each state's income threshold (Kaiser Family Foundation 2009). As SCHIP is distinct from Medicaid it provides a similar set of benefits to those of Medicaid recipients, but the benefits are not equivalent to the Medicaid benefits structure and patient copayment and coinsurance rates of some SCHIP recipients often more closely mirror a commercial health plan than the much lower out-of-pocket expenses encountered by Medicaid beneficiaries. The differences in benefit design should be accounted for in any study pooling SCHIP and Medicaid enrollees, particularly when payments are used as an outcome variable.

Of particular importance to the researcher using Medicaid claims is the issue of Medicare dual eligibles. As Medicare is the first or primary payer for these individuals, Medicaid pays for only a portion of health services not covered by Medicare such as a deductible, coinsurance, or for health services not covered at all by Medicare. This has a tremendous effect on the amount paid by Medicaid for services covered by Medicare, such as hospitalizations. As not all providers will bill Medicaid for the smaller portions of a bill not covered by Medicare, capture of all the services in the Medicaid claims file cannot be assured. Further complicating the use of data for the Medicare dual eligibles is that not all are entitled to full Medicaid benefits. Qualified Medicare beneficiaries (QMBs) and special low-income Medicare beneficiaries (SMBs or SLMBs) will have their Medicare Part B and or Part D premiums paid by Medicaid, but many are not eligible for typical Medicaid-covered services and these recipients should only rarely be included in research studies relying on Medicaid claims. One of the most profound changes for Medicare dual eligibles is the lack of a prescription benefit provided by Medicaid with the implementation of the Medicare Part D prescription benefit that went into effect January 1, 2006. With the rolling out of the Medicare Part D prescription benefit, nearly all measures of drug exposure are missing from the Medicaid prescription claims for Medicare dual eligibles. Essentially the Medicaid data from 2006 onward have no prescription information for elderly people or for many of the disabled recipients, which limits the scope of research for many retrospective drug studies, particularly studies focusing on drug therapies to treat chronic conditions such as cardiovascular diseases, arthritis, and diabetes, in which the preponderance are elderly or older people.

There are other beneficiaries and optional groups for which states may restrict the comprehensiveness of the Medicaid benefits, such as women eligible only for the family planning services (CMS 2005). Before researchers begin an analysis using any state Medicaid data, they should carefully examine the state's particular eligibility requirements and benefit structure for the recipients. Unless the research question is specifically addressing a narrow question related to recipients with partial Medicaid coverage, recipients not entitled to full Medicaid benefits should generally be excluded from analyses because there would be profound measurement bias due to the incomplete capture of health services, diagnoses, and prescription use for those individuals.

Medicaid covered services

Medicaid provides for health services across a fairly comprehensive range. State Medicaid programs must offer federally required services such as inpatient and outpatient hospital, laboratory, and physicians' services, and a range of other health services (CMS 2005). Notably, outpatient prescription drugs are not included as federally mandated services, but are optional services. All 50 states and the District of Columbia chose to offer an outpatient prescription benefit. Other optional services covered by most states include some form of dental coverage, eyeglasses, prosthetic devices, and physical and occupational therapy. It should be noted that not all beneficiaries are entitled to the entire benefits and services covered. Also, states possess the discretion to charge copays on most of the services, and may impose other restrictions such as precertification requirements or other usage restrictions limiting the maximum quantity of services per period of time (CMS 2005).

The National Pharmaceutical Council provides an excellent resource to learn about the pharmaceutical and some medical benefit designs across the 50 state Medicaid programs as well as the enrollment and spending summaries. It acquires this information through annual surveys of state Medicaid program administrators and the survey results are freely available.

Medicare claims data

Medicare, or title XVIII of the Social Security Act, was also enacted in 1965 to cover the healthcare services of over 45 million elderly and disabled people (CMS no date, b). The federal government, through the CMS, and not individual states, administers or oversees the Medicare program. Medicare benefits are tied to social security benefits and provide healthcare coverage to those aged >65, those who have been disabled for over 2 years, and those who have end-stage renal disease. The Medicare data are going to be the single best source for describing the healthcare service use of geriatric patients where most chronic diseases are prevalent because the preponderance of Medicare eligibles are elderly and nearly all legal residents in the USA are entitled to Medicare benefits.

Medicare provides coverage through three major parts: Part A (hospital insurance), Part B (physician and outpatient insurance), and Part D (outpatient prescription drug coverage). As noted, Part D coverage only came into being in 2006 and, before that, the Medicare program did not provide an outpatient prescription drug benefit; consequently, for pharmacists with an interest in studying the effects of drug exposure, only data after 2005 are of use. When people become eligible for Medicare, almost all are automatically eligible for Part A insurance; however, Part B and Part D coverage is optional and most Medicare recipients pay a monthly premium to access these benefits. Unlike Medicaid, Medicare coverage includes significant patient deductibles and coinsurance. For the researcher, it is essential that the study account for these elective Part B and Part D benefits, and include only recipients who enroll in these optional plans if the study requires measures derived from physician, outpatient services, or outpatient prescription benefits.

Medicare recipients can elect to enroll in the traditional fee-for-service Medicare plan or in a Medicare Advantage plan, sometime referred to as Part C. The Medicare Advantage plans are privately administered plans that utilize various health insurance structures such as preferred provider organizations (PPOs) or health maintenance organizations (HMOs) to provide benefits to recipients who elect coverage in a Medicare Advantage plan. Medicare recipients have their benefits paid by a commercial insurer so that their claims will not appear in the fee-for-service claims obtained through CMS, but their data may be acquired by obtaining the commercial insurance claims data from the insurer.

Medicare covered services

Part A of Medicare pays for inpatient care, some home health services including durable medical equipment, and limited skilled nursing facility services or rehabilitation services after a hospital stay of up to 120 days. Consequently, most long-term care in the USA is not paid for by Medicare and thus Medicare claims would substantially understate long-term care accessed by Medicare enrollees. Part B pays for physician and outpatient clinic services, ER, laboratory, ambulance, outpatient mental health services, and some preventive and wellness care services such as physical exams and diabetes wellness care. Most of the services are subject to coinsurance or copays. For people enrolled in traditional fee-for-service Medicare, the providers collect the recipient copays and deductibles, and bill CMS intermediaries directly.

The Part D Medicare benefits structure is fundamentally different from Parts A and B benefits. For Part D, Medicare enrollees sign up for a prescription drug plan (PDP) administered by a private third party payer such as a pharmaceutical benefit management (PBM) company or health insurer. The prescription drug plans offer services on a state-by-state basis, and Medicare recipients select among PDPs offered in their state. Each of the plans offers choices that vary in the comprehensiveness of their formulary, the monthly premiums, and patient copays. The benefit designs should be accounted for in research that seeks to compare drugs because the benefit structure will act as a powerful steering or channeling mechanism.

Acquiring Medicaid and Medicare claims data

Summary data for Medicaid and Medicare data are freely available from the CMS as public use files that describe aggregate enrollment and spending in individual states. These summary data, when used alone, have more limited research purposes but can be used to demonstrate cross-sectional relationships between states and various policies, as well as time trends, and are an important resource for describing the overall attributes of the programs. Additional nonidentifiable data on costs, such as Medicare cost reports of the Part B Extract Summary System (BESS) data file and the physician supplier procedure master file (PSPS), are important tools that are publicly available, which have particularly relevant uses in pharmacoeconomics to value health services (CMS no date, c).

Data describing Medicare claims are available in limited dataset (LDS) format as well as identifiable formats. Medicaid data obtained through CMS are available only as identifiable datasets. LDSs have all directly identifiable patient identifiers stripped from the claims but include elements of individual claims information (CMS no date, c). Data for Medicaid can generally be obtained from two sources: direct relationships with state Medicaid agencies or the CMS. Researchers can develop relationships with state Medicaid agencies and acquire the data from individual state agencies. Some states, such as California, have developed specific protocols for acquiring the data for Medi-Cal (California Medicaid), whereas most states do not have units designed specifically to make data available to researchers and access to the data is made on a case-by-case basis, usually starting with some collaborative relationship.

Medicare and Medicaid claims data can be obtained directly from the CMS for research purposes and additional arrangements can be made with individual state agencies. For Medicaid, the Medicaid Analytic Extract (MAX) files are research identifiable files that include dates of service and potentially identifiable beneficiary-level data elements. The data are organized in five files: the personal summary file including enrollment, demographic, and summary charge and expenditure data; the inpatient file; the long-term care file; the prescription drug file; and the other therapy file which contains claim records for professional services such as physician services, lab/radiograph, clinic services, and premium payments. Data can be requested for multiple states and the data are fairly affordable to the researcher. The price for acquiring MAX data from CMS for one to four states is US$1000–1500 for each year of data; however, these prices vary over time. One limitation with obtaining Medicaid claims from CMS is the 3- to 4-year lag in obtaining MAX claims records.

As Medicare is a federally administered program, the main source of Medicare claims data is through CMS. The Medicare claims data are organized similarly to the MAX files with individual files for inpatient, skilled nursing facility (SNF), home care, hospice, carrier (physician), drug (Part D drug event data), and durable medical equipment. MedPAR files may be used in place of the individual inpatient, SNF, and SNF stay files for Medicare recipients, but lack some of the details that would be available in the individual files. As the Part D benefit is administered by many Part D providers or pharmaceutical benefit management companies that have an array of formularies, copays, and doughnut hole coverage (where the recipient pays 100 percent of drug cost after they exceed a maximum coverage amount, until they reach the maximum out-of-pocket limit), a linkable file can be obtained describing the Part D plan characteristics.

Medicare data are not generally requested or made available on a state-by-state basis as for Medicaid, but rather are based on investigator inclusion or exclusion criteria or based on random samples such as a 5 percent sample. The price of obtaining Medicare data is generally more expensive than obtaining Medicaid data from the CMS and the prices of the individual files vary widely, are subject to change, and depend on the inclusion and exclusion criteria and the size of the sample requested. Medicare data are generally more up to date than Medicaid claims data with about a 2-year lag.

To acquire either Medicare or Medicaid data from CMS, a data use agreement (DUA) must be obtained. The DUA is for a specified set of research aims, but a data reuse request can be obtained to explore a new set of objectives for existing data already acquired, contingent upon CMS approval (CMS no date, c). The Research Data Assistance Center (ResDAC) is a CMS contractor that vets applications to acquire data from the CMS and provides free assistance to researchers interested in using Medicare and/or Medicaid data, in determining if and what CMS data are appropriate for the research, and in responding to questions about the data during analysis. The ResDAC website (www.resdac.umn.edu) contains the detailed instructions and forms to apply for a DUA and a flowchart describing the CMS DUA process is found in Figure 12.1. There are about a dozen documents that need to be completed as part of the DUA request process, and include an analysis plan or protocol for the project, institution review board (IRB) documentation, personnel, data management, and evidence of funding.

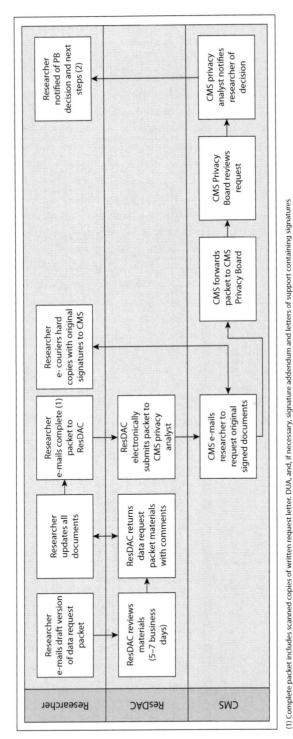

(1) Complete packet includes scanned copies of written request letter, DUA, and, if necessary, signature addendum and letters of support containing signatures
(2) A DUA cannot be assigned until the hard copies with original signatures are received

Figure 12.1 CMS data requisition submission process.

Typical administrative claims files structures

Generally administrative data are organized based on their source documentation or forms that they use to bill for services and the provider categories. Administrative claims data will generally fall into one of the following types of files: the enrollment or member eligibility file, inpatient and other facility files, professional (physician) service files, and the outpatient prescription files. Each of these categories has unique forms or source documents providing the structure governing the type of data in each of these files. Of course these categories can be further divided or subcategorized into smaller files based on the provider types of billing for those services within the larger category. For example, there may be separate files for inpatient acute care, long-term care, and outpatient facility charges. For each provider or category of service that is able to bill for services from a payer, such as Medicaid, the provider must meet the payer's requirements to be able to bill for a particular service. In the case of Medicaid and Medicare, provider manuals are available from the state Medicaid agency or those able to bill a particular state Medicaid program or Medicare. These provider manuals provide an excellent source for the researcher trying to identify services that are and are not covered, as well as the reporting requirements specific to each provider type (CMS 2009).

Over the last few decades paper submission of forms has been declining and many providers now submit claims electronically using a management information system, although paper claims may still be used. Nevertheless, the architecture of the paper claims and electronic claims use the same general standards for reporting the information. Once all the data are converted to electronic media for research purposes, there are no fundamental differences between paper and electronic claims.

To understand any research data, particularly for retrospective administrative data where the researcher is rarely involved with the data entry or the design of the data entry fields, an understanding of the source documents, forms, and templates is critical to guide the research process. This is particularly true for administrative data where the primary purpose of the data is not for research processes, but, rather, to pay providers and maintain current enrollment files to determine who is eligible for which benefits. Knowing what information is required for payment purposes and what is optional will have implications for determining the reliability of the information captured in the different fields. Those fields that are optional for payment purposes will be considerably less reliable than those that are required for payment purposes.

Administrative claims files

In most health claims records, there are two parts of a claim: a claim header and the claim details. The claim header records the patient and provider information, dates of service, as well as the summary billing information including the total amount billed, the amount collected from the patient, and information on the type of claim diagnostic such as an admission or discharge diagnosis. The claim details record the individual services delivered as part of an encounter such as a procedure performed or a drug that was dispensed. For each service or product delivered, there will be a unique detail record coupled with some of the header information. For more complex health claims such as an inpatient claim, there may be dozens of services billed in one claim (CMS 2009).

When a raw detail claims file is acquired it is imperative that the researcher recognize that each detail record does not represent a unique claim, but rather an individual detail of a service or product delivered that was billed as part of a claim. Summing across all detail records can grossly overstate utilization. An internal control number (ICN) is a number that is customarily assigned to each individual claim. It can be utilized to link header records with detailed records, and allows a researcher working with raw claims the ability to identify unique claims when working with detail records. Often the researcher will find that it is sufficient to use only the header information or the header records because that simplifies programming efforts. The header information often captures enough information, including the total amount billed and paid for all the services and products, to address many research questions. Also prescription claims files almost always have just one detail record for each header record so there is really not a meaningful distinction between headers and details for prescription claims (CMS 2009).

Once a health provider submits a claim to a health payer it can be paid, rejected by the payer, adjusted, or reversed by the provider originally filing the claim. A rejected claim can then be resubmitted for payment again, usually changing any incorrect or missing information; however, not all claim rejects can be resubmitted or paid. Once a claim has been paid, it will have the amount paid appended to the claim and will be considered an adjudicated paid claim (CMS 2009). Typically, researchers will use only paid claims in their analysis because they can have higher degrees of confidence that the services described in the claim are those that were actually delivered to recipients eligible to receive those services.

Depending on the source of administrative data, some claims may have been processed so that only final adjudicated paid claims are included and any adjustments to the claim have already been accounted for. Also some administrative claims may have the claims details "rolled up" in one record merged with the header information. Each record in the administrative data research files represents a paid claim and provides some details on the individual products or services delivered, but does not have separate records for each claim detail as might be encountered when working with raw claims (CMS 2009).

Inpatient and facility claims

Before 2008, the UB-92 form was the standard form to enter facility charges, including inpatient and institutional claims files. The UB-92 form was developed by the National Uniform Billing Committee (NUBC) and was the standard format used for reporting for acute care hospitals, hospices, psychiatric facilities, long-term care facilities, and hospices. In 2008, the UB-04 (HCFA-1450) file layout replaced the UB-92 as the common structure guiding submission of claims for facility charges (CMS 2009). A copy of the HCFA-1450 form is found in Figure 12.2. The areas that are cross-hatched are part of the header record. The area that is shaded represents one claim detail. If there are multiple details for a claim, a new line would be completed for each service provided and would represent a new detail record. Some important elements of the claims that are required to be complete when billing services for most health payers include the provider name and address (boxes 1 and 2), the type of bill (box 4), the dates of service (box 6), patient information (boxes 12–15), admission date (box 17), revenue codes (box 42), payer name (box 50), health plan identifiers (box 51), national provider identifier (box 56), principal diagnosis code (box 67), admitting diagnosis (box 69), and attending provider identifiers (box 76). It should be noted that secondary diagnoses are not required for billing purposes and, in the claim details, HCPCs (CPT-4) are used to describe the services billed. Coding conventions are outlined later in the chapter (CMS 2009).

Professional services and physician claims

The HCFA-1500 or the CMS-1500 form is the standard format for submitting health professional claims for physician services, clinics, home services, and other outpatient providers (CMS 2009). A copy of the form is found in Figure 12.3. The areas that are cross-hatched correspond to the header information and the area that is shaded corresponds to the claim details. Each line in box 24 would correspond to a claim detail that indicates a unique service as part of that professional encounter. As with the HCFA-1450 form, most of the elements, but not all, are required in order for the claim to be reimbursed by most providers. Key elements of the form that need to be completed are most of the patient information in boxes 1–8, insurance information (box 11), the diagnosis code(s) (box 21), the procedure or services performed entered separately on each line (box 24), the total charges for all the services delivered and the amount the patient paid (boxes 28 and 29), and the provider's information (boxes 32 and 33) (CMS 2009). Some noteworthy caveats regarding common elements needed are that not all providers are required to provide diagnoses. Researchers may consider restricting diagnostic information to include only codes

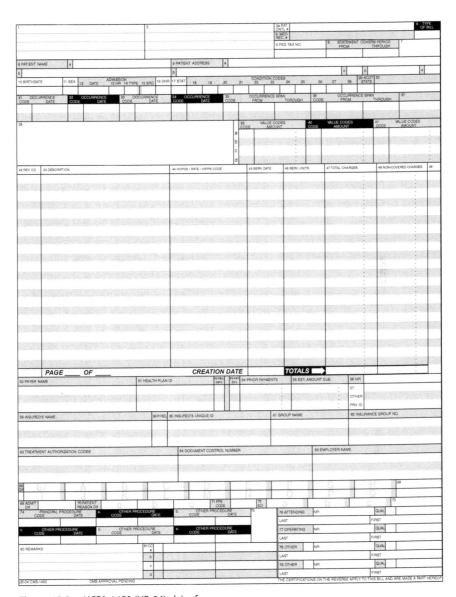

Figure 12.2 HCFA-1450 (UB-04) claim form.

obtained by provider types that would be expected to reliably diagnose disease such as physician, physician assistants, or nurse practitioners. The details also include diagnosis codes that correspond to each claim detail that may or may not be included in the diagnosis codes found in the header (box 21). Also each claim detail or service delivered in box 24 is coded with Healthcare Common Procedure Coding System (HCPCS) codes. It is important to note that the provider performing the service is often not the

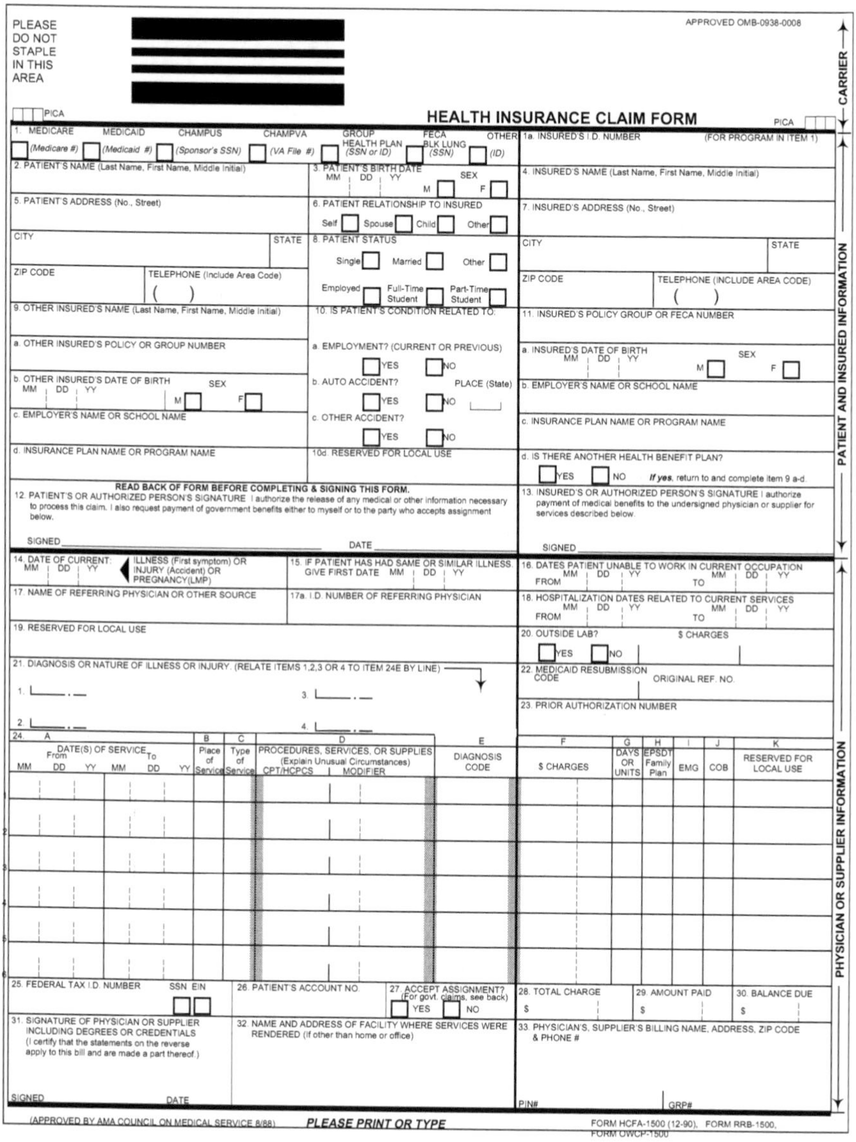

PLEASE DO NOT STAPLE IN THIS AREA

APPROVED OMB-0938-0008

CARRIER

PICA HEALTH INSURANCE CLAIM FORM PICA

1. MEDICARE MEDICAID CHAMPUS CHAMPVA GROUP HEALTH PLAN FECA BLK LUNG OTHER
(Medicare #) (Medicaid #) (Sponsor's SSN) (VA File #) (SSN or ID) (SSN) (ID)

1a. INSURED'S I.D. NUMBER (FOR PROGRAM IN ITEM 1)

2. PATIENT'S NAME (Last Name, First Name, Middle Initial)

3. PATIENT'S BIRTH DATE MM DD YY SEX M F

4. INSURED'S NAME (Last Name, First Name, Middle Initial)

5. PATIENT'S ADDRESS (No., Street)

6. PATIENT RELATIONSHIP TO INSURED Self Spouse Child Other

7. INSURED'S ADDRESS (No., Street)

CITY STATE

8. PATIENT STATUS Single Married Other; Employed Full-Time Student Part-Time Student

CITY STATE

ZIP CODE TELEPHONE (Include Area Code) ( )

ZIP CODE TELEPHONE (INCLUDE AREA CODE) ( )

9. OTHER INSURED'S NAME (Last Name, First Name, Middle Initial)

10. IS PATIENT'S CONDITION RELATED TO:

11. INSURED'S POLICY GROUP OR FECA NUMBER

a. OTHER INSURED'S POLICY OR GROUP NUMBER

a. EMPLOYMENT? (CURRENT OR PREVIOUS) YES NO

a. INSURED'S DATE OF BIRTH MM DD YY SEX M F

b. OTHER INSURED'S DATE OF BIRTH MM DD YY SEX M F

b. AUTO ACCIDENT? PLACE (State) YES NO

b. EMPLOYER'S NAME OR SCHOOL NAME

c. EMPLOYER'S NAME OR SCHOOL NAME

c. OTHER ACCIDENT? YES NO

c. INSURANCE PLAN NAME OR PROGRAM NAME

d. INSURANCE PLAN NAME OR PROGRAM NAME

10d. RESERVED FOR LOCAL USE

d. IS THERE ANOTHER HEALTH BENEFIT PLAN? YES NO *If yes*, return to and complete item 9 a-d.

READ BACK OF FORM BEFORE COMPLETING & SIGNING THIS FORM.

12. PATIENT'S OR AUTHORIZED PERSON'S SIGNATURE I authorize the release of any medical or other information necessary to process this claim. I also request payment of government benefits either to myself or to the party who accepts assignment below.

SIGNED DATE

13. INSURED'S OR AUTHORIZED PERSON'S SIGNATURE I authorize payment of medical benefits to the undersigned physician or supplier for services described below.

SIGNED

PATIENT AND INSURED INFORMATION

14. DATE OF CURRENT: MM DD YY ILLNESS (First symptom) OR INJURY (Accident) OR PREGNANCY(LMP)

15. IF PATIENT HAS HAD SAME OR SIMILAR ILLNESS. GIVE FIRST DATE MM DD YY

16. DATES PATIENT UNABLE TO WORK IN CURRENT OCCUPATION FROM MM DD YY TO MM DD YY

17. NAME OF REFERRING PHYSICIAN OR OTHER SOURCE

17a. I.D. NUMBER OF REFERRING PHYSICIAN

18. HOSPITALIZATION DATES RELATED TO CURRENT SERVICES FROM MM DD YY TO MM DD YY

19. RESERVED FOR LOCAL USE

20. OUTSIDE LAB? YES NO $ CHARGES

21. DIAGNOSIS OR NATURE OF ILLNESS OR INJURY. (RELATE ITEMS 1,2,3 OR 4 TO ITEM 24E BY LINE)

1. 2. 3. 4.

22. MEDICAID RESUBMISSION CODE ORIGINAL REF. NO.

23. PRIOR AUTHORIZATION NUMBER

24. A DATE(S) OF SERVICE From MM DD YY To MM DD YY	B Place of Service	C Type of Service	D PROCEDURES, SERVICES, OR SUPPLIES (Explain Unusual Circumstances) CPT/HCPCS MODIFIER	E DIAGNOSIS CODE	F $ CHARGES	G DAYS OR UNITS	H EPSDT Family Plan	I EMG	J COB	K RESERVED FOR LOCAL USE
1										
2										
3										
4										
5										
6										

25. FEDERAL TAX I.D. NUMBER SSN EIN

26. PATIENT'S ACCOUNT NO.

27. ACCEPT ASSIGNMENT? (For govt. claims, see back) YES NO

28. TOTAL CHARGE $

29. AMOUNT PAID $

30. BALANCE DUE $

31. SIGNATURE OF PHYSICIAN OR SUPPLIER INCLUDING DEGREES OR CREDENTIALS (I certify that the statements on the reverse apply to this bill and are made a part thereof.)

32. NAME AND ADDRESS OF FACILITY WHERE SERVICES WERE RENDERED (If other than home or office)

33. PHYSICIAN'S, SUPPLIER'S BILLING NAME, ADDRESS, ZIP CODE & PHONE #

SIGNED DATE

PIN# GRP#

PHYSICIAN OR SUPPLIER INFORMATION

(APPROVED BY AMA COUNCIL ON MEDICAL SERVICE 8/88) PLEASE PRINT OR TYPE FORM HCFA-1500 (12-90), FORM RRB-1500, FORM OWCP-1500

Figure 12.3 HCFA-1500 form.

one completing the claims information because these tasks are typically relegated to billing or coding personnel (CMS 2009).

Outpatient prescription claims

Claims submitted for outpatient drugs follow the standards of the National Council for Prescription Drug Programs, Inc. (NCPDP). The NCPDP is a not-for-profit standards development organization that develops and continually updates standards for the exchange of prescription information including standards guiding pharmacy payer transactions. Key elements of a pharmacy claim include information about the patient, the pharmacy provider, the drug product, and payment information. Drug products are identified by their national drug code (NDC). The NDC is an 11-digit code that follows the following convention:

xxxxx – yyyy – zz

- xxxxx designates the manufacturer and may have leading zeros or may omit the leading zeros
- yyyy designates the drug product and strength, specific to that manufacturer
- zz designates the packaging and packaging size for a particular product.

Each drug product will have a unique NDC code and, for products that are widely used and have multiple generic and brand manufacturers, there may be hundreds of NDC codes for the same chemical entity such as hydrocodone/acetaminophen. In addition, it should be noted that often the NDC is the only drug product information in a raw claims file. Other characteristics of the drug dispensed, such as the drug name, strength, or therapeutic class, would have to be merged with a drug dictionary, master file, or proprietary system, such as Redbook, to include these other descriptors with the drug claims files. Other data elements on prescription claims files include the quantity dispensed, the days supplied, and the dispensing date. The days supplied field is a pharmacist-entered field and its reliability cannot always be assumed, particularly for inhalers, prepackaged liquids, injectables, or topical products, which may pose significant measurement issues for adherence calculations. Provider and patient information similar to that found on the facility and professional services claims is also included.

In the absence of fraud, drug claims data are presumed to be very accurate with regard to drugs that were actually dispensed and paid for by a third party. Despite the relatively accurate measure of prescriptions that were paid for by some payer, prescriptions claims data cannot be presumed to provide a complete picture of all drugs used by a person eligible for benefits. For drugs that are not on a formulary, are available over the counter, have social stigma associated with their use, or have particularly high copays, people may purchase these with out-of-pocket monies or obtain samples from physicians, thus circumventing capture in the claims records. A prescription claim may also provide a false-positive measure of drug exposure if a drug was dispensed but never actually ingested. Despite these potential circumstances that may make prescription claims an inaccurate measure of drug exposure or use, empirical data suggest that claims data are fairly accurate with concordances of nearly 90 percent when compared with other sources (Martin and Cox 2008).

Medical coding conventions

Coding for morbidity

The *International Classification of Disease,* 9th revision, clinical modification (ICD-9-CM) codes are the standard by which diseases and some procedures are coded in the USA. ICD-10 codes, or the 10th revision, are used outside the USA and include causes of death, but are not used in most health claims available in the USA. ICD-9-CM codes were developed by the World Health Organization (WHO) and maintained in conjunction with the National Center for Health Care Statistics (NCHS) and the CMS. ICD-9-CM codes not only represent diseases, but also are a means for classifying injuries, symptoms, patient complaints, or procedures. ICD-9-CM codes are updated annually so the researcher should be aware that using the same set of ICD-9-CM codes may not provide consistent markers of disease over time. ICD-9-CM codes use the following format:

XXX.yy

- XXX: three-digit header for a disease cluster
- yy: fourth- and fifth-digit details

The three-digit headers group similar clusters of diseases or complaints, and are organized along body systems (CDC 2009). The fourth- and fifth-digit codes are not always available and, when more detail codes are available, they are generally required to be considered valid for billing purposes.

ICD-9-CM codes in the range 001.0–999.9 are generally used to identify the reason for admission/encounter and are used to classify diseases and injuries (e.g., infectious and parasitic diseases, neoplasm symptoms or signs, and ill-defined conditions). Signs and symptoms codes should be used only if the diagnosis cannot be determined and not added if they are common for the diagnosis. Probable or ruled-out diagnoses should generally be recorded; it is important to note that some codes do not necessarily indicate the presence of disease, particularly for conditions where there is a high degree of diagnostic uncertainty (CDC 2009).

There are also supplemental codes that use alpha characters instead of the numeric characters exclusively described above. The first set of supplementary codes is "V" codes used to identify factors that influence health status and contact with health services. These codes (V01.0–V89.09) are used to provide reasons for visits such as preventive services or newborn deliveries, and are considered valid codes for admission and discharge codes. Another set of supplementary codes are "E" codes which provide a means to code external causes of injury and poisoning and follow the E000–E999 format and range. These codes cannot be used generally or as admitting diagnosis and are used to supplement the cause of injury (CDC 2009).

Diagnostic-related groups

For inpatient encounters, diagnostic-related groups (DRGs) are used to categorize related hospital admissions for statistical or payment purposes. In 2007 there were 579 CMS DRGs, with a few more available for private payers, for pediatric and neonatal conditions. DRG codes follow an XXX format and are assigned using grouping software based on the principal diagnosis and may include up to eight additional diagnoses, age, gender, complications, discharge status, and procedures. DRG codes can be used to supplement ICD-9-CM codes to identify cases or people with a disease, but lack the specificity of ICD-9-CM codes (CMS no date, b).

Typical research issues of ICD-9-CM and DRG codes

The primary use of ICD-9-CM and DRG codes for retrospective researchers would be for the purposes of case ascertainment such as identifying people with prevalent or incident cases of disease that can be used to define the study participants, assess the comorbidity burden of participants across a range of conditions, or be used as a primary outcome measure. When an incident case of disease is used as an inclusion criterion to define study participants or one of the primary endpoints or dependent variables of a study, care should be taken to make sure that there is an adequate clean period – a time period when a person is eligible with benefits without having any codes or related codes for the condition. The longer the clean period, the greater the assurances that a new claim with a diagnosis represents an incident case rather than a diagnosis for an ongoing medical condition, which is particularly relevant for chronic diseases. The length of the clean period should be determined by the clinical nature of the disease and whether the condition is chronic or acute, the frequency of monitoring (coding) for the disease, the availability of data, and time definitions of past researchers. When ICD-9 or DRG codes are being used to quantify comorbidity or included as control covariates, differentiating incident from prevalent cases of disease is less critical and the researcher will not often employ a clean period to ensure incident cases (CDC 2009).

One of the critical challenges with the use of ICD-9 codes is handling the sheer number of codes. Identifying all the codes that could be mapped to a disease or condition can be especially challenging, particularly for conditions for which there is less diagnostic certainty or for broad complex disorders. Often there is no clear set of ICD-9-CM codes to use and there may be competing sets of case definitions used in the past. The researcher should recognize that there is a tradeoff between specificity (1 – false-positive rate) and sensitivity (1 – false-negative rate) between competing case definitions. The broader a case definition, the higher the sensitivity and the lower the specificity will be, and vice versa for more narrow case definitions (CDC 2009). For example, a case of asthma may be more broadly defined by one or more outpatient claims with an ICD-9-CM code for 493.**, or more narrowly defined by requiring two different outpatient claims with an ICD-9-CM code for 493.** separated by at least 1 day. Requiring only one outpatient claim will yield more cases of asthma and be more sensitive, but, as asthma diagnoses may be made in the outpatient setting, the broader definition may incorrectly misclassify some people with symptoms of asthma as having asthma and would therefore be less specific. Similarly relying on primary diagnoses may have higher specificity at the expense of sensitivity in defining cases. When there is significant uncertainty in defining cases, particularly for outcome measures or inclusion criteria, sensitivity analyses exploring several case definitions should be considered to determine the impact of alternative case definitions (Motheral and Fairman 1997). Case definitions should be based on previous research, ideally citing research that has validated a particular set of ICD-9 case definitions against external sources such as medical charts and other sources (Robinson et al. 1997; Hebert et al. 1999; Du et al. 2000, 2006; Walkup et al. 2000; Twiggs et al. 2002; Koroukian et al. 2003; Zhan et al. 2007).

If the investigator wishes to build a comorbidity measure or map ICD-9-CM codes or drug codes to clinically distinct classifications, there are several publicly available comorbidity indices with the most widely used being based on the Charlson index (Deyo et al. 1992; Clark et al. 1995; Iezzoni et al. 1997; Elixhauser et al. 1998; Lamers 1999; Kronick et al. 2000). Comorbidity indices can be used for two general purposes in retrospective studies. They serve as an initial basis to map ICD-9-CM codes into clinically distinct groups and these groupings can be used to build individual measures or variable indicators, and as separate variables in a regression analysis. Another use is that each of the comorbidity indices can be used to build a single metric of comorbidity where each category is weighted to form a single composite score. An alternative approach to map individual ICD-9-CM codes is to use the free online Clinical Classification Software (CCS), which maps all ICD-9-CM codes into hierarchical or nonhierarchical clinical groups available from the Agency for Healthcare Research and Quality (AHRQ).

Coding for services, procedures, and products

The HCPCS, commonly pronounced "hikpiks codes," is the initial basis for identifying services, procedures, supplies, and products in a medical facility. There are two levels of HCPCS codes. Level 1 codes are current procedural terminology (CPT-4), developed by the American Medical Association (AMA) to build a common listing of procedures and services performed primarily by physicians. Level 2 HCPCS codes describe ambulance services and durable medical equipment, prosthetics, orthotics, and supplies that do not have a CPT code. CPT codes use a 5-digit numeric coding structure and HCPCS codes use a 5-digit alphanumeric structure with the first digit corresponding to a letter followed by four numeric characters.

CPT codes can be grouped using the conventions found in Table 12.1. CPT codes may also include a modifier of two additional numeric digits to indicate that the service has been altered. For example, a CPT code with a modifier of "-62" would indicate that two surgeons were involved in the procedure. For providers reimbursed on a fee-for-service basis, there would be a fee schedule for each specific procedure and this is the primary basis for determining the amount paid for professional services. As CPT codes are a proprietary system maintained by the AMA, there are no widely available code dictionaries freely available, but one can obtain a CPT manual from the AMA and other sources to identify specific codes that may be of interest to a researcher.

Level 2 HCPCS codes are maintained by CMS in conjunction with AHRQ. Before 2003, local HCPCS codes (level 3) were used regionally to define services not specified in the first two HCPCS levels. However, the level 2 HCPCS codes have become more comprehensive, thus reducing the need for local codes. Researchers should, however, be aware that, the older the data they are working with, the more likely they may encounter local codes. Level 2 HCPCS codes are categorized based on the conventions described in Table 12.2. HCPCS code descriptors can be obtained freely from the CMS. Level 2 HCPCS codes can be particularly important for researchers wishing to study non-oral drugs that are routinely administered in a physician office or clinic and may have to become familiar with J-codes to identify use of many drugs administered intravenously or through intramuscular injections, including most biologics. HCPCS codes are generally going to be the only source to identify exposure to these drugs in a claims database setting.

Table 12.1 CPT code groups

Procedure type	CPT code ranges
Evaluation and management	99201 to 99499
Anesthesiology	00100 to 01999 and 99100 to 99140
Surgery	10040 to 69979
Radiology	70010 to 79999
Pathology	80002 to 89399

Table 12.2 HCPCS level 2 code groups

HCPCS type	HCPCS code ranges
Transportation	A0000–A0999
Medical supplies	A4000–A8999
Dental procedures	D0000–D9999
Durable medical equipment	E0100–E9999
Nonoral drugs administered	J0000–J9999
Orthotic/Prosthetic procedures	L0000–L9999
Vision/Hearing services	V0000–V5999

Summary and conclusions

This chapter provided an overview of the data sources that describe the health benefits delivered through the two largest public systems in the USA, Medicaid and Medicare. It also provided insights into common administrative data structure and coding conventions that can be applied to a variety of health claims data sources. These are important considerations in conducting pharmaceutical practice and policy research based on health claims data. However, there is an array of topics that were excluded from this chapter such as statistical approaches and human participant issues that are important and the reader should become familiar with these subjects.

As with any research endeavor, having a research hypothesis and clear operational definitions is going to strengthen the research and afford it more validity. Having predefined hypotheses, participant inclusion criteria, operational definitions, and an analysis plan is more critical in the field of observational claims research. Data are literally available for millions of participants at affordable prices; the potential to drive a particular finding by altering any of the steps in the research process is always present. There is simply no requirement to register the analysis plans of observational data analyses and a strong, ethical, hypothesis-driven approach to research is critical to obtain valid results. To that end, recently convened task forces have described good research practices for claims database research that highlight this critically important practice in observational research, among other valuable topics, covering the reporting, design, and analysis of retrospective data base research (Cox et al. 2009; Berger et al. 2009; Johnson et al. 2009).

Acknowledgments

Gary Moore or the University of Arkansas for Medical Sciences and Gerri Barosso and Faith Asper of ResDAC both graciously read a draft of this chapter in timely fashion and provided helpful comments.

Review questions

1 What demographic groups are most and least likely to be represented in a Medicaid database?

2 What demographic groups are most and least likely to be represented in a Medicare database?

3 Where would you find procedure codes in a claim: the header or the claim detail?

4 What are the standard codes to identify the following?

   a   Procedures

   b   Diseases

   c   Drugs

5 What is the standard form to bill for the following services?

   a   Physician

   b   Long-term care

   c   Inpatient hospital

References

Berger ML, Mamdani M, Atkins D, Johnson ML (2009). Good research practices for comparative effectiveness research: defining, reporting and interpreting nonrandomized studies of treatment effects using secondary data sources: The ISPOR Good Research Practices for Retrospective Database Analysis Task Force Report – Part I. *Value Health* **12**: 1044–52.

Centers for Disease Control (2009). *ICD-9-CM Official Guidelines for Coding and Reporting.* Online. Available at: www.cdc.gov/nchs/data/icd9/icdguide09.pdf (accessed October 11, 2009).

Centers for Medicare and Medicaid Services (2005). *Medicaid At-a-Glance 2005. CMS 2005.* Washington DC: US Department of Health and Human Services. Available at: www.cms.hhs.gov/MedicaidGenInfo/Downloads/MedicaidAtAGlance2005.pdf (accessed August 18, 2005).

Centers for Medicare and Medicaid Services (2009). *Medicare Claims Processing Manual Chapter 26: Completing and Processing Form CMS-1500 Data Set.* Washington DC: US Department of Health and Human Services. Available at: www.cms.hhs.gov/manuals/downloads/clm104c26.pdf (accessed September 10, 2009).

Centers for Medicare and Medicaid Services (no date, a). *Medicaid.* Washington DC: US Department of Health and Human Services. Available at: www.cms.hhs.gov/home/medicaid.asp (accessed October 11, 2009).

Centers for Medicare and Medicaid Services (no date, b). *Medicare.* Washington DC: US Department of Health and Human Services. Available at: www.cms.hhs.gov/home/medicare.asp (accessed October 11, 2009).

Centers for Medicare and Medicaid Services (no date, c). *Research, Statistics, Data & Systems.* Washington DC: US Department of Health and Human Services. Available at: www.cms.hhs.gov/home/rsds.asp (accessed October 11, 2009).

Clark DO, Von Korff M, Saunders K, Baluch WM, Simon GE (1995). A chronic disease score with empirically derived weights. *Med Care* **33**: 783–95.

Cox E, Martin BC, Van Staa T, Garbe E, Siebert U, Johnson ML (2009). Good research practices for comparative effectiveness research: approaches to mitigate bias and confounding in the design of nonrandomized studies of treatment effects using secondary data sources: The International Society for Pharmacoeconomics and Outcomes Research Good Research Practices for Retrospective Database Analysis Task Force Report – Part II. *Value Health* **12**: 1053–61.

Deyo RA, Cherkin DC, Ciol MA (1992). Adapting a clinical comorbidity index for use with ICD-9-CM administrative databases. *J Clin Epidemiol* 45: 613–19.

Du X, Freeman JL, Warren JL, Nattinger AB, Zhang D, Goodwin JS (2000). Accuracy and completeness of Medicare claims data for surgical treatment of breast cancer. *Med Care* 38: 719–27.

Du XL, Key CR, Dickie L, Darling R, Geraci JM, Zhang D (2006). External validation of Medicare claims for breast cancer chemotherapy compared with medical chart reviews. *Med Care* 44: 124–31.

Elixhauser A, Steiner C, Harris DR, Coffey RM (1998). Comorbidity measures for use with administrative data. *Med Care* 36: 8–27.

Hebert PL, Geiss LS, Tierney EF, Engelgau MM, Yawn BP, McBean AM (1999). Identifying persons with diabetes using Medicare claims data. *Am J Med Qual* 14: 270–7.

Iezzoni L, Ash AS, Daley J, Hughes JS, Schwartz M (1997). *Risk Adjustment for Measuring Healthcare Outcomes*, 2nd edn. Chicago: Health Administration Press.

Institute of Medicine of the National Academies (IOM) (2009). *Initial National Priorities for Comparative Effectiveness Research*. Available at: www.iom.edu/CMS/3809/63608/71025. aspx (accessed August 25, 2009).

Johnson ML, Crown W, Martin BC, Dormuth CR, Siebert U (2009). Good research practices for comparative effectiveness research: analytic methods to improve causal inference from nonrandomized studies of treatment effects using secondary data sources: The ISPOR Good Research Practices for Retrospective Database Analysis Task Force Report – Part III. *Value Health* 12: 1062–73.

Kaiser Family Foundation (2009). StateHealthFacts.org. Online. Available at: www.statehealth-facts.org/comparemaptable.jsp?cat=4&ind=204 (accessed August 18, 2009).

Koroukian SM, Cooper GS, Rimm AA (2003). Ability of Medicaid claims data to identify incident cases of breast cancer in the Ohio Medicaid population. *Health Serv Res* 38: 947–60.

Kronick R, Gilmer T, Dreyfus T, Lee L (2000). Improving health-based payment for Medicaid beneficiaries: CDPS. *Health Care Financ Rev* 21(3): 29–64.

Lamers LM (1999). Pharmacy cost groups: a risk-adjusted for capitation payments based on the use of prescribed drugs. *Medical Care* 37: 824–30.

Martin BC, Cox E (2008). The validity of electronic prescription claims data: A comparison of electronic prescription claims records with pharmacy provider records. *Value in Health* 11: A564.

Motheral BR, Fairman KA (1997). The use of claims databases for outcomes research: rationale, challenges, and strategies. *Clin Ther* 19: 346–66.

Robinson JR, Young TK, Roos LL, Gelskey DE (1997). Estimating the burden of disease. Comparing administrative data and self-reports. *Med Care* 35: 932–47.

Twiggs JE, Fifield J, Apter AJ, Jackson EA, Cushman RA (2002). Stratifying medical and pharmaceutical administrative claims as a method to identify pediatric asthma patients in a Medicaid managed care organization. *J Clin Epidemiol* 55: 938–44.

US Department of Health and Human Services (2009). Federal Coordinating Council on Comparative Effectiveness Research to Hold Public Listening Session in Washington DC. HHS Press Release, April 7, 2009. Agency for Healthcare Research and Quality. Available at: www.ahrq.gov/news/press/pr2009/hhscerpr.htm (accessed October 11, 2009).

Walkup JT, Boyer CA, Kellermann SL (2000). Reliability of Medicaid claims files for use in psychiatric diagnoses and service delivery. *Adm Policy Ment Health* 27: 129–39.

Zhan C, Battles J, Chiang YP, Hunt D (2007). The validity of ICD-9-CM codes in identifying postoperative deep vein thrombosis and pulmonary embolism. *Jt Comm J Qual Patient Saf* 33: 326–31.

Online resources

Agency for Healthcare Research and Quality. Healthcare Cost and Utilization Project (HCUP) Tools and Software. Available at: www.hcup-us.ahrq.gov/tools_software.jsp

The Centers for Medicare and Medicaid Services. HCPCS Release and Code Sets. Available at: www.cms.hhs.gov/hcpcsreleasecodesets/anhcpcs/List.asp

The Centers for Medicare and Medicaid Services. List of Diagnosis Related Groups. Available at: www.cms.hhs.gov/MedicareFeeforSvcPartsAB/Downloads/DRGdesc07.pdf

The Centers for Medicare and Medicaid Services. Medicaid Data Sources – General Information. Available at: www.cms.hhs.gov/MedicaidDataSourcesGenInfo/01_Overview.asp

The Centers for Medicare and Medicaid Services. Research, Statistics, Data & Systems. Available at: www.cms.hhs.gov/home/rsds.asp

The National Pharmaceutical Council (NPC). Medicaid Pharmaceutical Plan Resources. Available at: www.npcnow.org/Research.aspx

13

Secondary data analysis: commercial data

Robert J Valuck and Betsey Jackson

Chapter objectives

- To describe the major types of commercial data
- To present the key attributes of commercial data
- To discuss issues to be considered when choosing a commercial data source

Introduction

In the conduct of pharmaceutical practice and policy research, a number of research methods and data sources are available to investigators. When possible, randomized controlled trials and primary data collection are used, because they are considered the "gold standard" methods for establishing causal relationship. In some situations, however, it may be unethical, unfeasible, too time sensitive, or cost prohibitive to use such methods. In such cases, secondary data analysis is an alternate approach that can be employed to address a research question.

As described in Chapter 12, secondary data analysis involves the use of existing data (collected for purposes such as administration of healthcare insurance benefits, payment of provider claims for services provided) to conduct research. Secondary data may be obtained from various sources – government-sponsored health programs (Medicare, Medicaid, Veterans' Health Administration, etc.), national surveys (Medical Expenditure Panel Survey, National Health and Nutrition Examination Survey, etc.), registries (vital records, birth registries, the National Death Index), or from commercial vendors who obtain, aggregate, and de-identify the data and then make them available for a fee to researchers for the purposes of their investigations.

For the purposes of this chapter, commercial data are defined as patient-level data that are commercially available (i.e., can be licensed for research use), which does not necessarily imply that the patients (participants) in commercial datasets are covered by commercial insurance plans. Most commercial databases are derived from either administrative or clinical systems, available from vendors in the USA and, to a lesser extent, other countries around the world, reflecting the healthcare experiences of individuals living in those countries. Commercial databases offer researchers readily available, pre-existing sources of data for their work when other data sources (primary data collection, secondary data from government sources, etc.) are not viable or ideal.

This chapter is organized in an expanded-outline format, conducive to describing the major types of commercial data and their key attributes. This does not cover specific datasets from specific vendors in detail, because the marketplace for such datasets is ever changing and any highly detailed descriptions would fall out of date rapidly. Rather, for each type of commercial data, the following aspects are described:

- *Primary purpose of the data*: Why were the data collected?
- *Mechanisms of data provision*: How did the dataset become available?
- *Settings captured*: Which medical and/or pharmacy providers are included?
- *Data captured*: What data are captured in each setting? For example, pharmacy, diagnosis, procedures, charged and/or paid amounts, laboratory [automated/reference lab–not pathology in most cases], and/or clinical.
- *Systematic gaps*: Are there settings not captured?
- *Longitudinal or cross-sectional*: Can patients be followed over time? If so, how long are patients typically covered in the database?
- *Completeness*: Might some data be missing? For example, some prescription drugs may not be covered by insurance, including contraceptives or other "lifestyle" drugs.
- *Patient tracking in longitudinal data*: How do you know whether the patient is within the "window of observation"? Does no data mean no services (healthy) or that the data are not being captured?
- *Typical amount of history retained in a database*: How far back in time are claims/records available?
- *Typical database size*: How many patients are contained in the database?
- *Typical geographic representation*: Is the database local, regional, or national in scope?
- *Linkages*: Can the data be linked?
- *Research questions that can/cannot be addressed*: What are some of the typical research questions that can be answered using the database?
- *Examples of use*: What are some examples of the use of this type of database in pharmaceutical services or policy research?

Commercial data source types

There are two basic classes of commercially available secondary healthcare databases: administrative and clinical. Administrative databases originate from insurance billing and other financial functions, whereas clinical databases come from patient care delivery systems. The former tend to lack clinical detail and the latter financial (cost or payment) information.

The following subsections describe the basic types of databases within each of these two broad categories, using the framework described above. Although commercially available databases fall into one of the two categories, some vendors offer multiple types and, increasingly, linkages are being formed between various types of databases by single vendors and by vendor collaborations. An understanding of the essential database types and how they originate puts a researcher in a position to think through the more complex hybrid offerings and determine which databases are best suited for the researchers' needs.

Health Insurance Portability and Accountability Act (HIPAA) compliance, or individual privacy protection, is common to all databases and, of course, is taken very seriously. The point at which patient identifiers are encrypted or removed (thus de-identifying the data) varies and has implications for the feasibility and accuracy of linking data across sources. Research using commercial databases is most often conducted under the specific terms and conditions of a license agreement, with an additional data use agreement specifying additional details (such as who may have access to the data, how it will be transported and stored, how it will be returned or destroyed after the study). Although commercial database vendors may or may not require an institutional review board (IRB) review to be obtained by researchers licensing their data, it is a standard and accepted scientific practice that all research be reviewed and approved by an appropriate IRB before being undertaken.

Taking these issues into consideration, in practice no database type or source is the best; database selection is always a matter of finding the best fit between the research questions and the available databases. Some of the key considerations in finding the best fit are discussed later in the chapter.

Administrative databases

Insurance claims, open databases

These are very large databases with claims that are obtained from sources other than the claim payer.

- *Primary purpose of the data*: these data are collected in the reimbursement process for medical services or pharmacy purchases from health insurers to service providers. These are available only for individuals with some kind of health insurance (either public such as Medicare, Medicaid, and/or Veterans Affairs [VA]; or commercial/privately purchased such as from a Blue Cross and Blue Shield plan, Aetna, United, Humana, large self-insured employer). This is essentially billing data, derived from standard Centers for Medicare and Medicaid Services (CMS) billing forms; they have been universally adopted by payers.

- *How data become available*: datasets are available through acquisition from pharmacy benefit managers (PBMs), large pharmacy chains, medical providers, electronic data interchanges (clearing houses or "switches" which process claims from pharmacies, hospitals, and professional providers). Any of the steps in the claims processing cycle may be sources.
- *Settings captured*: primarily pharmacy claims for prescription drug purchases (not over the counter); also professional provider (both primary care and specialist), hospital outpatient, and inpatient setting are captured.
- *Data available from pharmacy claims*: the National Council for Prescription Drug Programs (NCPDP) standard data format is used for prescription drug claims. The fields extracted for secondary use usually include: national drug code (NDC) unique drug identifier, patient identifiers (may be encrypted from the source), patient demographics (usually not including ethnicity or race), quantity dispensed, therapy days supplied, patient copayment, amount billed to insurance, and actual amount paid by insurance. Data do not include proprietary or generic names, strength, or dosage; these can be linked based on the NDC code. Pharmacy data do not include diagnosis, because physicians are not required to provide that information when writing a prescription.
- *Data available from professional provider claims*: the extracted fields of CMS-1550 claim forms typically include patient identifiers (may be encrypted from the source), patient demographics (usually not including ethnicity or race), service date, provider identifier (may be encrypted from the source), provider specialty, place of service (can be inpatient), diagnosis, procedures, patient copayment, amount billed to insurance, and actual amount paid by insurance.
- *Data available from hospital claims*: the extracted fields of CMS-1450 (UB-04) claim forms typically include patient identifiers (may be encrypted from the source), patient demographics (usually not including ethnicity or race), service date, admission date if applicable, provider identifier (may be encrypted from the source), diagnoses, procedures, patient copayment, amount billed to insurance, and actual amount paid by insurance. Hospital claims do not include medications administered at the facility. Hospital medications are billed as a lump sum for drugs.
- *Systematic gaps*: claims that are processed outside the catchment of the particular database are not captured.

- *Longitudinal vs cross-sectional*: longitudinal. It is possible to observe a patient over time; typically, patients can be followed for 2 or more years.
- *Completeness*: open claims databases do not necessarily capture all claims for a given individual, because a person may use varying providers and not all of those will be captured. Patient eligibility for insurance is not captured, so a lack of claims may represent either a lack of insurance or a period of good health with no services. For example, in a study of healthy young people experiencing repeat ear infections, lack of a second antibiotic claim or physician encounter could result from lack of repeat treatment, lack of capture of services due to changes in providers, or lack of capture due to a change in insurance status. Research in chronic conditions claim activity is a proxy for presence. As all-payer databases, these capture claims paid by public payers (Medicare, etc.).
- *Patient tracking*: the lack of insurance eligibility data makes the total period of patient presence in a database, start and end dates, unknown.
- *Typical amount of history retained in a database*: most datasets involve 3 or more years, and more than 5 years for some data.
- *Typical database size*: these datasets are very large containing significant percentages of all electronic claims processed.
- *Typical geographic representation*: most datasets are nationally representative.
- *Linkages*: claims data can be linked with data from automated laboratories in some cases.
- *Research questions that can/cannot be addressed*: questions related to healthcare product/service utilization studies, patterns of care, and trend analysis can be addressed. Some retrospective observational designs can address comparative effectiveness, safety, and cost. No prospective or randomized designs are possible, so, for comparative studies, advanced statistical methods must be used to address issues of nonrandom allocation, channeling bias, confounding by indication, etc. Further, population sizes (denominators) typically are not available, so calculation of true incidence/prevalence rates may not be possible.
- *Examples of use*: Segal et al. (2007) used commercial data from i3Innovus to estimate the effects of long-term treatments on total healthcare charges and hospitalizations in patients with new diabetes medications versus established drugs. In this example, new diabetes drugs showed no demonstrable differences in outcomes relative to existing therapies. Robinson et al. (2006) used the Medstat MarketScan Commercial Claims and Encounter database to study rates of adherence to National Committee for Quality Assurance (NCQA) medication management measures for antidepressant treatment. They found that only 19 percent of depressed patients started on antidepressants met all three NCQA treatment appropriateness criteria. Libby et al. (2007) used the

IMS/PharMetrics Patient Centric Database to study the effects of the US Food and Drug Administration (FDA) Black Box Warning on antidepressants and suicidality on patterns of depression care among US managed care enrollees. They found that, after the warning, rates of diagnosis and pharmacologic treatment declined, reversing several years of prior growth trends.

Insurance claims, closed population databases

These are very large databases with claims and insurance eligibility data that are obtained from claim payers, typically from many payers:

- *Primary purpose of the data*: same as in open claims data above.
- *How data become available*: directly or indirectly from health insurers. These databases are sometimes called "managed care databases"; however, not all insurance plans are managed care plans. Some insurers license their own data for research; although concerns about member sensitivity are a barrier, some conduct research using their own claims data but will not release the data. Data are available indirectly through businesses providing services to health plans (as do large self-insured employers) and through pooling data from numerous clients in the process; the pooled data become a secondary revenue stream.
- *Settings captured*: pharmacy (retail and some mail service), professional provider (both primary care and specialist), and hospital (outpatient and inpatient facility services).
- *Data available within each setting*: same as in open claims data above.
- *Systematic gaps*: any medical services that are not paid by the health insurer.
- *Longitudinal vs cross-sectional*: longitudinal. It is possible to observe a patient over time, typically for periods of 1–2 years, as a function of people changing employers and/or health plans.
- *Completeness*: any medical services that are not paid by the health insurer will be missing. This includes "lifestyle" drugs, some contraceptives, some nontraditional therapies, and 100 percent Medicare-paid services for elderly people. Some do not capture mail service drugs for chronic conditions. Some lack behavioral health services provided by third party providers. Note that each health plan has a unique benefits structure; benefits information is not generally available.
- *Patient tracking*: insurance eligibility data are included in these databases, making it clear when the individual enters and exits the database and whether coverage was for medical, pharmacy, or both services at any point in time. Eligibility data also include type of plan (managed care, preferred provider organization, etc.). Accurate tracking is possible.

- *Typical amount of history retained in a database*: 5 or more years.
- *Typical database size*: in any given recent year, 10–20 million total eligible individuals.
- *Typical geographic representation*: national.
- *Linkages*: with data from automated laboratories; hospital administrative data including inpatient medications.
- *Research questions that can/cannot be addressed*: similar questions can be answered, and approaches used, as were noted previously for "open" insurance claims databases.
- *Examples of use:* Valuck et al. (2003) used data from Aetna, Inc. to identify demographic, clinical, and pharmacotherapy-related factors associated with response to lipid-lowering drug therapy among members of a US managed care organization. They found that, despite better lipid-lowering results among users of higher potency hydroxymethylglutaryl coenzyme A (HMG-CoA) reductase inhibitor (statins) drug regimens, only a small percentage of participants received such regimens. Nair et al. (2005) used data from a single western US managed healthcare plan to evaluate the impact of increased copayments on the continuation of nonformulary medications in multitiered pharmacy benefit plans. They reported that individuals confronted with increased copayments often switched their nonformulary drugs to formulary-listed alternatives. McClure et al. (2007) used data from Kaiser Permanente Health Plan of Colorado to study the risk of myositis associated with statin and fibrate drug use, and found that these drugs were significantly predictive of increased myositis risk, along with renal and liver disease as other independent risk factors.

Hospital databases, administrative

These are databases of billing records obtained from hospitals or hospital systems:

- *Primary purpose of the data*: billing, either insurers or patients, and for internal administration such as cost accounting.
- *How data become available*: through hospital collaborations or companies that provide analytic services to hospital clients and have the right to license the data for research.
- *Settings captured*: varies; all include inpatient (IP) data (defined as at least one night stay); some also include outpatient (OP) services such as emergency room, outpatient day surgeries.
- *Data available for IP, OP*: patient demographics (may include ethnicity); drug administered (not prescriptions given at discharge); all or most

diagnoses; all or most procedures; cost or cost-to-charge ratios. Some are beginning to include hospital laboratory results from the hospital labs.

- *Systematic gaps*: no care outside the hospital, so it is not possible to follow, for example, drugs used pre- and post-discharge.
- *Longitudinal vs cross-sectional*: cross-sectional data. It is not possible to track patients across hospitals. Same-hospital care can be tracked.
- *Completeness*: complete; all data are captured for the encounter because these are "all payers" databases.
- *Patient tracking*: an individual returning to the same hospital or hospital group can be identified as the same patient, otherwise tracking is not possible.
- *Typical amount of history retained in a database*: at least 5 years.
- *Typical database size*: the unit of measure for these databases is the discharge because patients cannot be identified uniquely. These databases have 20–40 million total discharges.
- *Typical geographic representation*: quite broad; a function of the contributor base; some can project to the USA.
- *Linkages*: with claims data to create longitudinal data; all-service data with inpatient drug data.
- *Research questions that can/cannot be addressed*: hospital admissions and/or discharge studies; patterns of hospital care; cost of hospital care or diagnosis-related group (DRG) studies. May be difficult to study the use of specific medications and utilization patterns, or to study care before or after hospital stays.
- *Example of use*: Kieszak et al. (1999) used hospital billing data to determine whether administrative data could be used to replicate the Charlson Comorbidity Index (originally developed via manual review of medical charts on a subject-by-subject basis). They reported that the index was found to be a significant predictor of inpatient mortality, 30-day mortality, length of stay, and complications, after controlling for age, gender, and neurologic and medical risk factors.

Clinical databases

Electronic health records databases, hospital

These are databases of patients' electronic health records (electronic charts; EHRs) obtained from hospitals or hospital systems.

- *Primary purpose of the data*: hospital clinical systems to deliver patient care.
- *How data become available*: typically from companies that sell the systems and make arrangements for data sharing among users.

- *Settings captured*: hospital IP, possibly hospital OP services.
- *Data available*: hospital demographics: number of beds, geographic region, type of hospital (teaching, nonteaching, specialty, etc.); patient demographics, including ethnicity; insurance information; medications given (usually from the hospital pharmacy system); diagnoses; procedures; laboratory results including microbiology; billed charges (not paid amounts).
- *Systematic gaps*: no care that was provided outside the hospital, so it is not possible to follow, for example, drugs used pre- and post-discharge.
- *Longitudinal vs cross-sectional*: cross-sectional data. It is not possible to track patients across hospitals. Same-hospital care can be tracked.
- *Completeness*: complete, in that all data are captured for the encounter because these are "all payer" databases.
- *Typical amount of history retained in a database*: 5 or more years.
- *Typical database size*: 10+ million discharges.
- *Typical geographic representation*: quite broad; a function of the contributor base.
- *Linkages*: potential, particularly with claims data.
- *Research questions that can/cannot be addressed*: hospital admissions and/or discharge studies; patterns of hospital care. May be difficult or impossible to study costs, or the use of specific medications and utilization patterns, or to study care before or after hospital stays.
- *Examples of use*: Hunteman et al. (2009) used data from an inpatient hospital to analyze rates of drug-allergy alerts and responses among medications ordered in the hospital's computerized charting system, and found that, although only 1.3 percent of orders generated drug-allergy alerts, most of the alerts were overridden by prescribers. Hemstreet et al. (2006) used hospital laboratory and medical records data to evaluate potassium and phosphorus repletion in hospitalized patients, and to assess the role of computerized decision support systems for improving the use of oral electrolyte replacement therapy dosing. They found that computerized decision support systems can reduce rates of inappropriate potassium and phosphorus use in the hospital setting.

EHR databases, physician

These are databases of patients' EHRs (electronic charts) obtained from individual physicians or practices.

- *Primary purpose of the data*: clinical systems to deliver patient care, i.e., "the automated or computer-based chart."

- *How data become available*: typically from companies that sell the systems to physicians and make arrangements for data sharing among users.
- *Settings captured*: predominantly primary care community-based physicians with practices of varying sizes, some specialists.
- *Data available*: practice demographics, patient demographics including ethnicity, patient history for key conditions, insurance information, diagnoses, any procedures, blood pressure, height, weight, smoking status, laboratory results for many common tests (HbA1c, blood glucose measures, etc.). Medication orders are included but not prescriptions filled.
- *Systematic gaps*: no other settings are captured, e.g., a hospitalization would not be recorded.
- *Longitudinal vs cross-sectional*: longitudinal. Patients may be followed as long as they remain within the practice (this is typically 2–3 years, sometimes longer).
- *Completeness*: complete, in that all data are captured for the encounter because these are "all payer" databases.
- *Patient tracking*: a patient is presumed to be within the window of observation as a function of medical activity – a long period of absence suggests leaving the practice.
- *Typical amount of history retained in a database*: 5 or more years.
- *Typical database size (total patients)*: can be millions of patients.
- *Typical geographic representation*: quite broad; a function of the contributor base.
- *Linkages*: potential, particularly with claims data.
- *Research questions that can/cannot be addressed*: physician practice patterns (patterns of diagnosis, treatment, procedures, referrals, etc.); physician prescribing patterns (vs dispensed drug patterns in pharmacy claims); studies needing laboratory data (orders, results), clinical data (family history, social history), or other more specific detailed clinical data. May not be possible to study costs of care, or prescribed versus dispensed medication use. Observational study designs will require advanced statistical approaches to address nonrandom allocation, confounding by indication, etc.
- *Example of use*: Pace et al. (2009a) describe the use of a network of EHR data from eight organizations (representing more than 500 clinicians and 400 000 patients) to conduct comparative effectiveness and safety research that is enhanced with clinical information not available in traditional claims-based analyses. A report (Pace et al. 2009b) from their study of comparative effectiveness and safety of oral diabetes medications (ODMs) for adults with type 2 diabetes mellitus found no demonstrable differences across individual ODMs in terms of effectiveness or safety.

EHR databases, integrated delivery network

These are databases of patients' EHRs (electronic charts) obtained from networks of physicians, practices, or clinics:

- *Primary purpose of the data*: clinical systems to deliver patient care across the entire medical delivery network. Geisinger Health Systems and Henry Ford Health System are examples of integrated delivery networks (IDNs).
- *How data become available*: either directly from a single IDN or from collaborations among multiple IDNs or other groups coordinating data sharing.
- *Settings captured*: all physicians (primary care and specialists), hospitals, laboratories, and in some cases pharmacies within the IDN. Retail pharmacy activity is typically not included.
- *Data available*: patient demographics including ethnicity, patient history for key conditions, insurance information, diagnoses, any procedures, blood pressure, height, weight, smoking status, laboratory results for many common tests (HbA1c, blood glucose measures, etc.), as well as specialized tests such as bone mineral density and spirometry in some cases. Medication orders are included but not prescriptions filled.
- *Systematic gaps*: services provided outside the IDN are not captured, such as visits to a non-network specialist or hospital.
- *Longitudinal vs cross-sectional*: longitudinal. Patients can be tracked from first to last encounter; typically this is over a span of a few years, sometimes many years – again, depending on patient turnover or movement from one healthcare system to another.
- *Completeness*: all in-network services are included. Some clinical detail recorded in text fields (notes) may not be available.
- *Patient tracking*: lack of activity could mean a true lack of activity, treatment outside the system, or patient has left the network.
- *Typical amount of history retained in a database*: 5 years.
- *Typical database size*: individual IDNs tend to be in the range of <500 000 total active patients.
- *Typical geographic representation*: narrow for individual IDNs, broader but specific for multiple IDN collaborations.
- *Linkages*: with claims data for the subset of individuals who are both treated in the IDN and insured by the associated insurer.
- *Research questions that can/cannot be addressed*: similar questions can be answered, and approaches used, as were noted previously for physician-level EHR databases.
- *Examples of use*: Newman et al. (2000) described the implementation of an osteoporosis disease state management program in the Penn State Geisinger Health System, and found that the multidimensional program resulted in improvements in osteoporosis outcome and process measures

across this large, rural health network. McCullough et al. (2002) used data from the Henry Ford Medical System in Detroit, Michigan, to confirm an increased rate of heart failure diagnoses in that area, corresponding with epidemiologic studies from other regions of the USA.

Therapeutic area-specific (or disease-specific) clinical databases

These are databases of individuals with a specific condition or disease, or taking a certain medication, and are used for longitudinal follow-up studies:

- *Primary purpose of the data*: tailored clinical systems for patient care in specialized clinics such as oncology or HIV.
- *How data become available*: typically from companies that sell the systems and make arrangements for data sharing among users.
- *Settings captured*: specialists, group practices, and clinics treating the condition.
- *Data available within each setting*: similar to other clinical systems, varies by system.
- *Systematic gaps*: no other settings are captured, e.g., hospitalization or treatment for other conditions would not be recorded.
- *Longitudinal vs cross-sectional*: patients can be tracked as long as they are treated at the site.
- *Completeness*: most treatment within the clinic is captured, although the research data are an extract and may lack important information, especially that recorded in text fields such as notes.
- *Patient tracking*: patients can be tracked within the clinic; generally a patient's reason for disappearance from the data is not known.
- *Typical amount of history retained in a database*: 2 or 3 years; can be as many as 10.
- *Typical database size*: between 100 000 and 300 000 total patients.
- *Typical geographic representation*: specific regions where the systems have been implemented.
- *Linkages*: potential, particularly with claims data.
- *Research questions that can/cannot be addressed*: descriptive (case report or case series) studies of clinical presentation, patterns of treatment (including drug utilization and dosing detail), outcomes, and costs for patients with specific medical conditions, if data are available. Historical control studies are possible, as are registry-type analyses. Randomized or comparative observational studies are difficult.
- *Examples of use*: Davis et al. (1995) used data from the Coronary Artery Surgery Study (CASS) registry to compare 15-year survival rates among men and women receiving either medical or surgical management of coronary artery disease. They reported that those receiving initial surgical

treatment survived longer, particularly those at high risk. Piccirillo et al. (2004) used a hospital-based cancer registry to assess the effect of comorbidities on survival for patients with cancer. They reported that comorbidity is an important, independent prognostic risk factor for patients with cancer, and recommended that comorbidity information be routinely collected in disease-based registries.

Considerations in choosing among commercial data sources

Deciding whether a commercial database may be most suitable for a particular research study or question (and, if so, which one) is a complex matter, and many factors must be taken into account, including an understanding of the key attributes of the major database types (covered above), as well as the specific nuances of a particular research study, the design being employed, and the question being addressed.

To illustrate some of the many nuances that must be considered when choosing among commercial data sources, a study referenced previously (Pace et al. 2009b) is used as a case example. In that study, Pace and colleagues were funded by the federal government to evaluate the comparative effectiveness and safety of ODMs for adults with type 2 diabetes mellitus. In performing the first part of their study, the research team faced a decision about which commercial data source to use, to help them establish background rates and patterns of ODM use, and relative rates of effectiveness (reduction in glycated hemoglobin or HbA1c) and safety (occurrences of liver enzyme elevation or liver failure) of ODMs in primary care. Issues that had to be considered in selecting a commercial database for this purpose included:

- *Continuous observation*: often longitudinal studies such as the retrospective cohort design employed in this case require lengthy continuous observation periods. Pace et al. needed to identify new cases of diabetes, so they required a 1-year period of eligibility for each participant without any claims indicative of a diabetes diagnosis or an ODM prescription. They also required at least 1 year of follow-up after ODM initiation, to allow sufficient time for outcomes to possibly occur and be detected.
- *Hospital detail*: sometimes studies focus on the use of medications in the hospital setting; in the case example, no such information was necessary as the focus was on outpatient treatment in primary care.
- *Clinical detail*: in this case, specific parameters such as body mass index (BMI) and family history of diabetes were considered "wish list" variables (i.e., not required, but desirable) in the commercial data source.

- *Time period*: continuous eligibility of at least 2 years was required for the ODM study. A 5-year time window (i.e., 2003–2007) was chosen, to allow for the identification/accrual of sufficient numbers of study participants.
- *Laboratory requirements*: lab results required data elements to establish the primary effectiveness and safety outcomes, so a database with lab results was needed. Further, participants had to have both a baseline HbA1c and at least one follow-up HbA1c test result 1 month or more after starting ODM therapy, to be included in the study population.
- *Recentness of the data*: the study period (5-year time window) was also defined more specifically as the most recent 5 calendar years of data availability (i.e., 2003–2007), with the goal of reflecting current treatment patterns as closely as possible. Thus, participants meeting the study inclusion criteria, but starting ODM before 2003, were excluded.
- *Patients >65 years of age*: the study population was adults with type 2 diabetes, and there was no specific need to study participants aged >65 years. Some were included in the database, but this was not by design for this particular study. Had this been a specific need, other database(s) might need to have been considered.
- *Financial information*: cost analysis was not a focus of the study; but, if it were, attention would need to have been given to the extent and nature of the financial information contained in the database (e.g., drug costs, medical costs, laboratory costs).
- *Financial detail*: there was no cost analysis; had there been such a focus, data elements such as patient coinsurance amount, copayment amount, deductible amount, payer paid amount, and provider submitted amount may have been needed.
- *Mortality*: mortality was not an outcome focus. Generally, mortality data themselves are not available; "discharge status" is sometimes available, but is an incomplete proxy.
- *Productivity*: productivity was not included in this study as an outcome or intermediary variable focus. Had that been the focus, a database that includes variables such as sick time, workers' compensation claims, and disability claims may have been desirable so that indirect cost analysis could be performed.
- *Budget*: the budget for acquisition of a commercial data source was sufficient to purchase an extract of data and obtain a 1-year license for analysis of the data. As of this writing, most commercial databases are available for single-study license fees of US$50 000–120 000 with varying conditions regarding term of use.
- *Analytic capacity*: the research team was experienced in using, handling, analyzing, and publishing the results from commercial data sources.

They used large computer servers at a university to house the data and run the analyses, and to ensure that privacy and confidentiality were protected (per the grant terms, and the database license terms). Experienced programmers and analysts were employed to run the analyses.

In the case example, several commercial databases were considered for possible use, and a database from Ingenix (the Impact National Managed Care Database) was used for the first stage of the analyses. Subsequently, EHR data from a network of primary care physicians (DARTNet) were used in second stage analyses, to replicate the initial findings and augment them with detailed clinical data (Pace et al. 2009a, 2009b).

Finally, although it is impossible to list all of the commercially available databases that might be considered for research purposes, several database resources and catalogues are available that describe a number of these databases. It is worth consulting these catalogues, reviewing them, and specifying questions to the database vendors directly or to individuals who have expertise in this area. The International Society for Pharmacoeconomics and Outcomes Research (ISPOR) has developed an electronic index (Digest) of key attributes of healthcare databases around the world – 192 in total, 90 in the USA, and 102 in other countries as of July 2010 – and the digest is available at the ISPOR website, which is accessible to the public. The digest is grouped by country and allows key word searches and searches by type of database, and can be accessed at the URL provided at the end of this chapter (under Online resources).

Summary and conclusions

Commercial data sources – both administrative databases originating from insurance billing and other financial functions, and clinical databases derived from patient care delivery systems – offer an important and viable source of data for many research questions and designs. Secondary data analyses using commercial data have proven to be an important part of the outcomes researcher's armamentarium, particularly: when large populations receiving standard care in the community are of interest; when prospective, randomized studies are not possible, ethical, feasible, or affordable; or when the question being addressed is of particular urgency. Many commercial databases are available, each having its own unique strengths and limitations, and researchers should consider these databases when formulating their research questions, study designs, and analysis plans. Specialized database agents may also be available to assist researchers with determining available data options and the resource(s) that best fit their needs.

Review topics

1 Describe the major types of commercial databases available to researchers, focusing on sources of data (i.e., origin of the data, or where generated).
2 List key attributes of commercial databases, comparing and contrasting the data that come from administrative claims versus electronic health records.
3 Discuss some of the issues that must be considered when identifying/ selecting the best data source for a particular research question/project.

References

Davis KB, Chaitman B, Ryan T, *et al.* (1995). Comparison of 15-year survival for men and women after initial medical or surgical treatment for coronary artery disease: A CASS registry study. *J Am Coll Cardiol* 25: 1000–9.

Hemstreet BA, Stolpman N, Badesch DB, *et al.* (2006). Potassium and phosphorus repletion in hospitalized patients: implications for clinical practice and the potential use of healthcare information technology to improve prescribing and patient safety. *Curr Med Res Opin* 22: 2449–55.

Hunteman L, Ward L, Read D, *et al.* (2009). Analysis of allergy alerts within a computerized prescriber-order-entry system. *Am J Health-Syst Pharm* 66: 373–7.

Kieszak SM, Flanders WD, Kosinski AS, *et al.* (1999). A comparison of the Charlson Comorbidity Index derived from medical record data and administrative billing data. *J Clin Epidemiol* 52: 137–42.

Libby AM, Brent DA, Morrato EH, *et al.* (2007). Decline in treatment of pediatric depression after FDA advisory on risk of suicidality with SSRIs. *Am J Psychiatry* 164: 884–91.

McClure DL, Valuck RJ, Glanz M, *et al.* (2007). Statin and statin-fibrate use was significantly associated with increased myositis risk in a managed care population. *J Clin Epidemiol* 60: 812–18.

McCullough PA, Philbin EF, Spertus JA, *et al.* (2002). Confirmation of a heart failure epidemic: findings from the Resource Utilization Among Congestive Heart Failure (REACH) study. *J Am Coll Cardiol* 39: 60–9.

Nair KV, Valuck RJ, Allen RR, *et al.* (2005). Impact of increased copayments on the discontinuation/switching rates of nonformulary medications. *J Pharm Technol* 21: 137–43.

Newman ED, Starkey RH, Ayoub WT, *et al.* (2000). Osteoporosis disease management: best practices from the Penn State Geisinger Health System. *J Clin Outcomes Manage* 7: 23–8.

Pace W, West D, Valuck R, *et al.* (2009a). Distributed Ambulatory Research in Therapeutics Network (DARTNet): summary report. Prepared by University of Colorado DEcIDE Center under contract no. HHSA29020050037I-TO2. Rockville, MD: Agency for Healthcare Research and Quality. July 2009. Available at: effectivehealthcare.ahrq.gov/reports/final.cfm (accessed 15 September, 2009).

Pace WD, Cifuentes M, Valuck RJ, *et al.* (2009b). An electronic practice-based network for observational comparative effectiveness research. *Ann Intern Med* 151: 1–3.

Piccirillo JF, Tierney RM, Costas I, *et al.* (2004). Prognostic importance of comorbidity in a hospital-based cancer registry. *JAMA* 291: 2441–7.

Robinson RL, Long SR, Chang S, *et al.* (2006). Higher costs and therapeutic factors associated with adherence to NCQA HEDIS antidepressant medication management measures: analysis of administrative claims. *J Manag Care Pharm* 12: 43–54.

Segal JB, Griswold M, Achy-Brou A, *et al.* (2007). Using propensity scores subclassification to estimate effects of longitudinal treatments: An example using a new diabetes medication. *Med Care* 45: S149–57.

Valuck RJ, Williams SA, MacArthur M, *et al.* (2003). A retrospective cohort study of correlates of response to pharmacologic therapy for hyperlipidemia in members of a managed care organization. *Clin Ther* 25: 2936–57.

Online resources

General Practice Research Database (GPRD) Group. "The database." Available at: www.gprd.com/products/database.asp.

IMS. "IMS Health Data Offerings." Available at: www.imshealth.com/portal/site/imshealth/menuitem.a953aef4d73d1ecd88f611019418c22a/? vgnextoid=987e20ca11280210VgnVCM100000ed152ca2RCRD&vgnextfmt=default.

Ingenix/i3. "Data Assets." Available at: www.i3global.com/DataAssets.

International Society for Pharmacoeconomics and Outcomes Research (ISPOR). "ISPOR International Digest of Databases." Available at: www.ispor.org/Intl_Databases/index.asp.

Thomson Reuters. "Healthcare Research Solutions." Available at: www.thomsonreuters.com/products_services/healthcare/healthcare_products/research.

14

Secondary data analysis: national sample data

Rajender R Aparasu

Chapter objectives

- To describe national sample data sources
- To explain the complex sampling scheme in national surveys
- To present analytical considerations for national survey data analysis
- To discuss research approaches involving national survey data
- To identify strengths and weaknesses of national surveys

Introduction

National surveys are sponsored by various federal and nonfederal agencies to provide valuable healthcare data at the national level. Surveys conducted by federal agencies are primarily designed to provide data to congress and policy makers. Public use data files from national surveys are available to researchers and practitioners to address various pharmaceutical practice and policy-related issues. The National Center for Health Statistics (NCHS) is the principal federal data collection agency under the Centers for Disease Control and Prevention (CDC). Other federal agencies conducting national surveys include the National Institutes of Health (NIH), the Centers for Medicare and Medicaid Services (CMS), and the Agency for Healthcare Research and Quality (AHRQ). National survey data are also available from the researchers funded by nonfederal agencies, such as the Robert Wood Johnson Foundation or the Kaiser Family Foundation.

Secondary data from national surveys are valuable because it is relatively easy to obtain the data and the findings are generalizable to the nation. There are a number of national survey data sources that are relevant to pharmaceutical practice and policy research. Most of the surveys involving medication use data are conducted by the federal agencies, such as the NCHS, the AHRQ, and the CMS. This chapter primarily presents common national surveys conducted by federal agencies involving medication use data and explores the complex sampling scheme involved in these national surveys. The chapter also discusses common research approaches involving national survey data and the analytical considerations due to the complex sampling scheme. Finally, the strengths and weaknesses of national surveys are identified.

National surveys

The NCHS conducts a number of annual and periodic surveys at the population and provider levels (Kovar 1989). Population-based surveys collect person- or patient-level data using interviews and examinations. The unit of observation in population-based surveys is usually the person or patient. Examples of population-based surveys include the National Health Interview Survey (NHIS) and the National Health and Nutrition Examination Survey (NHANES). Provider-based surveys collect utilization data from various clinical settings using surveys and data abstraction procedures. The unit of observation in provider-based surveys is generally the patient visit or discharge. Examples of provider-based surveys include the National Ambulatory Medical Care Survey (NAMCS), the National Hospital Ambulatory Medical Care Survey (NHAMCS), and the National Nursing Home Survey (NNHS). The population-based surveys provided by other agencies include the Medical Expenditure Panel Survey (MEPS) and the Medicare Current Beneficiary Survey (MCBS). The data from these surveys are available to researchers and practitioners and brief descriptions are provided below.

National Health Interview Survey

The NHIS collects detailed health status and healthcare utilization data on the US civilian noninstitutionalized population (Botman et al. 2000). A nationally representative probability sample of approximately 40 000 households and noninstitutional group quarters participate in the NHIS. Computer-assisted personal interviews (CAPIs) are used by trained professionals to capture data on approximately 100 000 individuals in the households. The NHIS is an annual cross-sectional survey based on stratified multistage sample design that involves a probability sample of primary sampling units (PSUs), area segments, and permit segments within the PSUs, households within the segments, and finally the respondents within the households. In national surveys, a PSU is usually a county, a few adjacent counties, towns or township, or a metropolitan area. The NHIS obtains person-level data on chronic conditions, healthcare utilization, health status, health insurance, access to medical care, health behaviors, and immunization. The basic or core module and other interview supplements are used to collect data at various levels, namely household, family, and individual or person levels.

The basic module contains the family core, sample child core, and sample adult core. The family core collects family-level information such as family size and income; the sample child core collects detailed health status, health service, and health behavior data on one child per family; the sample adult core collects similar detailed data relevant to one adult per family. The other interview supplements vary from year to year depending on the current health topics. The data files provided for secondary data analyses include household files, family and person-level files, sample adult files, and sample child files. Person-level files can be merged with family- and household-level files to obtain compressive data. This survey can provide detailed patient-level data on morbidity, health-related disability, and healthcare utilization. Medication-related data collected include prescription drug benefit, injury due to medications, complementary medicine, and medication use by children. The NHIS does not provide nature and extent of use of prescription medication. The analysis can be performed at household and family levels to evaluate costs and access issues. The sampled adult and child data can be used to conduct analysis at the population level. The NHIS data are often used in addressing the burden of a disease, access to prescription medications, and prescription benefit issues.

National Health and Nutrition Examination Survey

The NHANES collects detailed health and nutrition data on the US civilian noninstitutionalized population from interviews, physical examinations, and laboratory tests (National Center for Health Statistics 1994). The NHANES is a continuous, multi-year cross-sectional survey based on a complex, four-stage probability cluster design with a sampling scheme similar to the NHIS. Multiple data collection methodologies such as interviews, physical examinations, and laboratory tests are used to collect data from approximately 5000 individuals each year. The NHANES uses CAPI to collect extensive data on health conditions, environmental exposures, health behaviors, and nutritional supplements and prescribed medications usage. With respect to prescription medications, all medications used by individuals in the past 30 days are collected using a medication inventory method.

The NHANES uses mobile examination centers (MECs) to conduct physical exams, physiologic measurements, and laboratory tests. The exams include vision, audio, oral, and body measurements. The laboratory component involves collection of biological and environmental specimens for various diagnostic tests. These exams are conducted by trained professionals. Not all respondents participate in the medical exam component. The multiple data files are provided for secondary data analysis for each of the following components: demographic, examination, laboratory, and questionnaire. All files can be merged using a respondent sequence number. Person-level data from the NHANES can be used to examine disease or medication prevalence patterns based on responded–reported data. The NHANES is often used for determining the nature and extent of prescription medication use in noninstitutionalized populations. The data source can also be used to address prescription medication quality and access issues across diseases. The analytical sample usually includes data collected for 2–4 continuous years for sample size considerations.

National Ambulatory Medical Care Survey

The NAMCS collects detailed data on provision and use of ambulatory services rendered in nonfederal physician offices in the USA (Bryant and Shimizu 1988). The NAMCS is a cross-sectional annual survey of in-person office visits. The NCHS utilizes a multistage probability design that involves probability samples of PSUs, physician practices within the sampling units, and patient visits within those practices. Over 1500 participating office-based physicians complete the patient records for a systematic random sample of their office visits during the assigned reporting period. Data are reported by physicians or their office staff at the visit level using patient record forms. The patient record form includes information pertaining to patient demographics, reasons for visit, physician diagnoses, diagnostic services provided or ordered, drugs prescribed, and disposition of the visit. Physician characteristics such as specialty and region of practice are also collected before the survey implementation.

Prescribing data involve up to eight prescription and nonprescription medications, including all new or continued medications ordered, supplied, or administered during each visit. Dosage and duration are not captured in the NAMCS. Limited clinical data are collected in the NAMCS, such as blood pressure, temperature, and body mass index. Laboratory tests ordered by physicians are recorded but the test results are not captured. The NAMCS data are available as a single analytic file for each year. The physician-collected data from the NAMCS can be used to examine disease-specific visit rates, use of ambulatory resources, and drug-specific or disease-specific utilization reviews. The unit of observation and analysis is the visit and NAMCS data cannot be used to examine population-based prevalence estimates. The NAMCS is useful in addressing quality-of-care issues, such as evidence-based and inappropriate use of medications. Prescribing trends can also be examined by combining multiyear data.

National Hospital Ambulatory Medical Care Survey

The NHAMCS provides data on ambulatory care rendered in hospital emergency rooms and outpatient departments (McCaig and McLemore 1994). The NHAMCS is a cross-sectional annual survey based on samples of patient visit records from the emergency rooms and outpatient departments of noninstitutional general and short-stay hospitals. The survey utilizes a four-stage sampling design to select the PSUs, hospitals within a PSU, departments or clinics within hospitals, and finally visits within the selected departments or clinics. Data describing the clinical nature of the emergency visits include the patient's problem or complaint, prior visit status, referral status, physician's diagnosis, visit-related injury, surgical procedures performed, urgency of the visit, diagnostic and screening services, therapeutic services, counseling/advice provided, medications rendered, disposition of the visit, and the providers seen.

The data collection instrument in outpatient departments is similar to that of the NAMCS. Consequently, the data from the NAMCS and outpatient departments of the NHAMCS can be combined to examine prescribing patterns in nonemergency ambulatory settings. Physician specialty data are not collected in the NHAMCS; instead it identifies the type of outpatient department. Two analytical files are available for each year to provide data on emergency rooms and outpatient departments. The emergency visit data from the NHAMCS can be used to examine emergency visits and medication use patterns in emergency rooms. The NAMCS and NHAMCS can be combined to examine resource utilization patterns including medication use across ambulatory settings. The combined NAMCS and NHAMCS data can also be used to address access and quality-of-care issues in ambulatory settings.

National Nursing Home Survey

The NNHS is designed to provide nationally representative information on nursing homes from two aspects: provider of services and recipient of care in the USA (Jones et al. 2009). It is a periodically conducted cross-sectional survey. The sampling is based on a stratified two-stage probability design, with the first stage involving facilities and the second stage residents within facilities. Over 1000 nursing homes participate in the first stage by providing facility information on over 13 000 residents. The NNHS is conducted using a CAPI. It contains two facility-level modules and four resident-level modules that are completed for up to 12 residents per facility. The facility-level modules are completed by an interviewer before the resident-level modules to confirm the eligibility of the facility for the survey. Resident-level data are organized into four modules: health status, nonminimum dataset, prescribed medications, and payment sources. The data files for the NNHS include facility, resident, and prescription files.

The facility file contains facility characteristics, such as facility size, facility ownership, Medicare/Medicaid certification, types of services provided, staffing patterns, and specialty programs offered. The resident file contains resident's demographic characteristics, health status, diagnoses, services received, charges, and sources of payment. The prescription file includes up to 25 medications administered within 24 hours and up to 15 medications taken by the resident on a regular basis, but not taken in the last 24 hours. Similar to the NAMCS, dosage and duration of medication are not captured in the NNHS. Although the NNHS is a provider-based survey, resident and prescription files can be combined to analyze medication use patterns at the resident level. Facility-level data cannot be combined with other files for confidentiality issues. The NNHS data can be used to assess nature and extent of prescription use in nursing home settings in the United States.

Medical Expenditure Panel Survey

The MEPS is sponsored by the AHRQ and NCHS to provide the most comprehensive information on healthcare use, expenditures, payment, and coverage for the noninstitutionalized population in the USA (Cohen 1997). The MEPS utilizes a multistage, clustered, sampling design to collect data from a nationally representative sample of the US civilian noninstitutionalized population. The federal agencies started conducting the MEPS annually in 1996. The multiple overlapping panels or cohorts of respondents for the MEPS are derived from the NHIS to provide annual utilization and expenditure data. The data collection for the MEPS involves four components: household, medical provider, insurance, and nursing home. The medical provider and insurance components are based on the household component, which serves as the core survey for the MEPS. The household component collects medical utilization and expenditure data at both individual and household levels. Specifically, detailed data are collected on demographics, heath conditions, health status, use of medical care including medications, charges and payment, access to care, insurance coverage, income, and employment. The medical provider component validates medical care utilization data by contacting medical providers and pharmacies identified by household respondents. The providers include hospitals, physicians, home health agencies, and pharmacies. The insurance component collects information on health insurance plans from employers, unions, and other sources of insurance. The nursing home component is a survey of nursing home residents. The insurance and nursing home components are not available as public use data files for confidentiality reasons.

In the household component, five rounds of in-person interviews are conducted over a 30-month data collection period to capture person level use and expenditure data for two calendar years. A combination of data collection techniques including computer-assisted in-person interviews, telephone interviews, and mailed surveys are used throughout the year. Each interview collects utilization data for 3–4 months for all household members. The medical provider component collects detailed data on the medical and financial characteristics of services received by respondents. The AHRQ releases the annual full-year consolidated data files from household and medical provider components. The annual files provide complete data on all respondents participating in one particular year of the MEPS. The event files capture claims-type utilization data on hospitalization, outpatient visits, emergency visits, prescription utilization, and use of home healthcare services. Only month and year of the healthcare utilization event are available for confidentiality reasons. As each panel provides data for two calendar years, longitudinal studies can be conducted by combining annual consolidated and event-level files for each year based on respondent identifiers. The MEPS data can be used to evaluate a broad range of issues pertaining to cost, access, and quality of pharmaceutical care.

Medicare Current Beneficiary Survey

The MCBS is a continuous nationwide sample survey of aged, disabled, and institutionalized Medicare beneficiaries (Adler 1994; Sharma et al. 2001). The MCBS is sponsored by the CMS to collect detailed sociodemographic, health status, utilization, and expenditure data on Medicare users. The MCBS is a longitudinal survey based on data collected from a multistage stratified random sample of Medicare beneficiaries. The stratification is by age within older and disabled subgroups. Four-year rotating panels are used for the MCBS. A sample of approximately 12 000 beneficiaries is interviewed annually using CAPI. Approximately 16 000 beneficiaries are available from the MCBS due to rotating panel design. Each panel is interviewed, three rounds per year for 4 years, to capture the longitudinal data on Medicare beneficiaries. Respondents are interviewed irrespective of residence using appropriate community or long-term questionnaires. Interviews are conducted to collect data on healthcare utilization and expenditures, health status, and functioning.

With respect to prescription medications, data are collected on all new and refilled prescriptions, including strength and number of doses. The respondent data are supplemented with administrative claims data. Baseline and community core questionnaire instruments are used for beneficiary data collection. The facility interview includes facility core and baseline questionnaires. The data from the surveys are used to prepare annual access to care, and cost-and-use files. Access to care files contains data on access to healthcare, satisfaction with care, and usual source of care, and other relevant information, supplemented by claims data. The cost-and-use files combine claims and survey data to provide complete data on utilization and expenditures from all sources. The files include all medical events, including use of prescription medications. The MCBS is a valuable data source to conduct cross-sectional and longitudinal studies to address various costs, access, and quality of pharmaceutical care issues in elderly people.

National survey design and analysis

Sampling design

National surveys conducted by the federal agencies are based on a complex sampling design involving multistage sampling, clustering, stratification, and disproportionate sampling (Lee et al. 1989; Korn and Graubard 1999). These techniques improve operational efficiency of data collection and provide nationally representative data to address various policy-relevant issues. Multistage sampling in national surveys involves multiple stages of selection of sampling to minimize costs and time for data collection. The first stage of selection is usually the largest sampling unit, also known as the PSU. This usually involves cluster sampling to select a probability sample of PSUs that contain a group or a cluster of basic sampling units (Cochran 1977; Thompson 2002). In national surveys, the first stage sample selects PSUs defined as a county, a few adjacent counties, towns or township, or a metropolitan area. These are mutually exclusive and exhaustive geographic areas. According to the NCHS, there are approximately 1900 geographically defined PSUs that cover all states and the District of Columbia.

In the NAMCS, the first stage cluster sampling consists of 112 PSUs randomly selected from the available 1900 PSUs (Bryant and Shimizu 1988). The second stage involves selection of a stratified sample of physician practices within the PSUs. Stratified sampling is designed to comprehensively represent sampling units from each stratum level either for comparative reasons or for subgroup analyses. In the NAMCS, all physician specialties are grouped into 15 specialty groups, and selection of practices is made within each specialty group for a sufficient representation of specialties in the final sample. Most national surveys involve disproportionate stratified sampling to select sampling units with a higher probability of selection from within the small subpopulation or stratum compared with strata with high subpopulations. This is usually done to provide sufficient samples for underrepresented populations, such as black and Asian individuals, to conduct subgroup analyses.

The probability of selection in a national survey is not equal due to disproportionate sampling. This unequal probability of selection has to be reflected in the survey data analyses. Therefore, survey sampling weights are assigned to survey respondents to adjust for different sampling probabilities in national surveys (Korn and Graubard 1999; Department of Economic and Social

Affairs 2005). In the NAMCS, the survey sampling weights adjust for probability of selection of PSU, probability of selection of practice within PSU, and patient visit within the practice. Inverse probability weighting is often used to account for probability of selection. For example, in a particular practice, if the probability of selection of visits in the final stage is 1 in 4 (or 0.25), then each visit within that practice is assigned a survey weight of 1 over 0.25 or 4. The value of survey sampling weight for a given visit in the NAMCS can be interpreted as number of visits in the population with the same or similar visit characteristics. In addition to probabilities of selection, survey weights are also used for nonresponse bias, fixed totals, and weight smoothing. Consequently, the survey design characteristics and survey sampling weights have to be incorporated in the statistical analysis to obtain valid national estimates.

Sampling error estimations

Sampling error is an error in estimate based on the sample rather than the population (Cochran 1977; Thompson 2002). It is one of the important measures of variability in the analysis of data. An estimator of sampling error is used to provide accurate variability of measures based on the samples. Sampling errors estimations in national surveys are complicated due to complex sampling designs. Although complex sampling designs increase operational efficiency, analyses of national survey data are computationally complex. Sampling techniques, such as clustering and stratification, influence variance and standard error calculations. Stratified sampling reduces sampling variation due to selection of similar sampling units within the stratum. Cluster sampling increases sampling variation due to selection of dissimilar sampling units. Consequently, statistical analyses that do not account for sampling design characteristics provide erroneous variance and standard errors. This can lead to flawed statistical significance. Most standard statistical procedures assume simple random sampling and hence are not suitable for complex survey data. Estimates and standard errors derived from such analyses are biased. The ratio of standard error adjusted for complex design sample, and standard error calculated assuming simple random sample is called the design effect (Kish 1965; Korn and Graubard 1999). It is also defined as the variation inflation factor. This ratio influences statistical significance and the conclusions drawn from the statistical tests.

The two most commonly used methods to estimate standard errors and variance in national survey data analysis are the Taylor series linearization and the replication methods (Wolker 1985; Korn and Graubard 1999). The Taylor series linearization method requires full design information for calculation of a linear approximation of variance that can be used for standard error and confidence interval estimations. The calculation of variance is based on a weighted combination of variance across PSUs within the same stratum. Estimates from each of the strata are used to provide overall variance estimation. This involves use of the first-order or linear part of the Taylor series expansion with the assumption that the contribution of higher-order terms is negligible. Taylor series linearization involves a separate variance estimate calculation for each type of estimator. Most NCHS surveys provide first-stage cluster and stratum variables for calculating variance based on the Taylor series linearization method.

Replication methods calculate variation based on estimates from the full sample and repeated subsamples or replicates from the full sample (Wolker 1985; Korn and Graubard 1999). Overall, estimated variance is based on the deviation of the replicate estimates from the full sample. These methods use weighted replicates for variance estimation and do not require full design information for the variance estimate calculation. Balanced repeated replication and the jackknife replication methods are frequently used replication methods. The former involves selection of the replicate by randomly sampling one PSU per stratum and is often used for multistage stratified sampling with at least two PSUs per stratum; in the latter, one cluster is removed in a multistage cluster sampling design to create a replicate. Both the Taylor series linearization and the replication methods provide comparable variable estimates for measures involving smoothed data such as means, proportions, or totals. However, replication methods, such as balanced repeated replication, are preferred for nonsmoothed functions such as medians (Korn and Graubard 1999).

Statistical software packages

The NCHS provides complex sampling design variables or weighted replicates for calculating variance based on the Taylor series linearization and the replication methods. However, the former is the commonly used analysis of data based on nation surveys because it is easy to implement and the design variables are often included in the national survey data. Recent national surveys have included only PSU, stratum, and survey weight variables. Replication weights are not needed for the Taylor series linearization method. Standard procedures in statistical packages cannot implement the Taylor series linearization method to analyze national survey data. SAS, STATA, and SUDAAN are the commonly used statistical packages to implement the Taylor series linearization method for analyzing national survey data. SPSS, Epi-Info, WesVar, PC-CARP, and IVEware are also available to analyze national survey data. The complete original survey structure has to be maintained for all analyses, including subgroup analyses. The design structure is needed to implement Taylor series linearization methods for variance estimations. The domain analysis option is often used to conduct subgroup analyses in the statistical packages.

SAS survey statistical procedures can implement the Taylor series linearization and the replication methods to analyze national survey data (SAS

Institute 2006). Most names of such survey procedures begin with SURVEY. The descriptive procedures include SURVEYFREQ and SURVEYMEANS. The regression procedures use SURVEYREG and SURVEYLOGISTIC. The GENMOD, MIXED, and NLMIXED procedures also have the capability of analyzing complex survey data. The design variables can be incorporated into these procedures using the keywords STRATUM, CLUSTER, and WEIGHT. Subgroup analyses can be performed using the DOMAIN option in these statistical procedures. SAS survey procedures are part of the standard statistical package and do not require special purchase.

SUDAAN is a statistical software package specifically developed to analyze complex survey data (Research Triangle Institute 2005). It can be used as a stand-alone package or as an SAS-callable routine. The descriptive procedures include DESCRIPT, CROSSTAB, and RATIO. The regression analyses use REGRESS, LOGISTIC, MULTILOG, and SURVIVAL procedures. The design variables can be incorporated into these procedures using the keywords NEST and WEIGHT. Subgroup analyses can be performed using the SUBPOPN option with these statistical procedures. SUDAAN can implement both the Taylor series linearization and the replication methods. It can also incorporate more complex sampling schemes than SAS. However, stand-alone packages have very limited data manipulation capability, so there is a need to use other software packages for data management purposes. SUDAAN is not a multipurpose statistical software and has limited capability to analyze noncorrelated data.

STATA is a multipurpose statistical software with the capability of implementing the Taylor series linearization and the replication methods to analyze national survey data (StataCorp. 2007). Most names of such survey procedures begin with SVY. Descriptive procedures include SVYMEAN, SVYTOTAL, SVYPROP, and SVYTAB. Regression procedures include SVYREG, SVYLOGIT, SVYOLOG, and SVYMLOG. The recent STATA has incorporated survey analysis capabilities in several other statistical procedures, including survival analysis and instrument variable regression. The design variables can be incorporated in these procedures using the keywords STRATA, PSU, and PWEIGHT. Subgroup analyses can be performed using the SUBPOPN option. STATA can manage and manipulate large datasets and can also be programmed to implement user-defined analyses.

Other software packages, such as SPSS, Epi-Info, WesVar, PC-CARP, and IVEware, can also be used to analyze complex survey data. The SPSS survey module is not a part of the standard package and has to be purchased separately. Epi-Info software, developed by the CDC, is freely available and has the capability to conduct descriptive analyses. WesVAR can be used for descriptive and regression analyses that involve the replication method for variance estimation. PC-CARP is limited capability software developed by Iowa State University to analyze complex survey data. IVEware, developed by the University of Michigan, is freely available SAS-callable software that can perform descriptive and regression analyses.

Research considerations

The NCHS surveys provide valuable data to conduct pharmaceutical practice and policy research. The NCHS-sponsored surveys can be downloaded freely from the NCHS website. These public use data files do not contain any identifiers and can be used only for statistical reporting. Federal law prohibits use of data for identification purposes. The NCHS datasets are available to researchers in several formats, including SAS and STATA. The NCHS datasets are coded and labeled appropriately with extensive documentation. The documentation provides valuable information on survey design, data collection, variable definition, and coding. The documentation is the primary source of information for the survey data. Although most of the data collected from the surveys are available to researchers in public use data files, data collected from the surveys that might jeopardize privacy and confidentiality, such as geographic or genetic data, are not released. However, such data are available to researchers at the NCHS Research Data Center upon request. The restricted data access requests have to be evaluated by the NCHS staff and restricted data files can be accessed remotely or at the Research Data Center in Hyattsville, MD.

Research question and design

The national surveys can be used to address a variety of cost, quality, and access issues related to medication use. The documentation is valuable in evaluating the utility of the national survey data for addressing the research question. Research studies involving the NCHS data are mostly cross-sectional because the designs used to conduct surveys are cross-sectional. Only the MCBS and the MEPS can provide longitudinal data. The MCBS data files, however, lack specific dates of healthcare utilization. The MEPS provides month and year of utilization, but not the day due to confidentiality issues. Outcomes research involving longitudinal designs, such as cohort and case–control, are not generally used for research based on national survey data. However, time series analyses involving longitudinal monthly utilization data can be performed. As the public use data files do not contain any direct identifiers, the research conducted using the NCHS data is usually considered exempt from review by the full committee of the institutional review board (IRB). The research proposal involving the NCHS data should be submitted to the IRB to receive the exempt status.

Sample size considerations are complex in research involving national survey data. The minimum sample size needed for analysis is based on the type of measure, such as mean or proportion, relative standard error, design effect, and degrees of freedom. Most of the sample size calculations for descriptive analyses are based on the relative standard error, which is defined as estimate divided by its standard error and is usually expressed as a percentage. The maximum acceptable relative standard error for most surveys is 30 percent. Most survey documentations provide the minimum sample needed for subgroup analyses. These considerations should be made for each of the subgroups of interest. Some strategies used to increase the sample size include use of multiyear data and utilization of limited exclusion criteria. Operational definitions of key variables should be the same when combining multiyear data.

Data extraction and analysis

Most simple survey datasets are provided as a single file; other extensive national surveys provide multiple data files. The data for each respondent can be combined across multiple files using a respondent number. The unit of observation may differ from the unit of analysis. In the NAMCS, the office visit is the unit of observation as well as the unit of analysis. In the MEPS, data are collected at event level and person level. The event-level data, such as hospitalization, need to be converted to the person level to analyze the data. The study objective defines the subgroup of interest. Missing data do not pose a significant problem in survey data analysis because they are very limited or imputed by the NCHS. Secondary data analyses usually require recoding and redefining of variables based on the needs of analyses. Usually this requires extensive use of data management programs such as SAS or STATA.

Most survey data analyses are focused on a subgroup of interest (i.e., children, adults, or elderly people). It is important not to subset the data in order to maintain the integrity of the original survey structure of the Taylor series linearization method. The domain analysis option in statistical packages can be used to conduct subgroup analyses. The data analyses considerations are driven by study objective and hypothesis. Novice researchers have a tendency to conduct extensive statistical analyses to explore all possible associations due to availability of numerous variables in national surveys. Researchers can overcome this by focusing the analyses on the subgroup of interest and limiting statistical tests relevant to the hypothesis under consideration. Frequency distribution and means with a design-adjusted confidence interval can help to describe the study sample. As the survey data are based on the sample rather than the population, confidence intervals are important in providing the variability of the estimates. Consequently, software packages, such as SAS, STATA, and SUDAAN, are needed to provide the design-adjusted confidence intervals.

Most studies involve some form of regression analysis to analyze national survey data. The dependent variable of interest can be indicators of cost, access, or quality of pharmaceutical care. The dependent variable can be constructed or modified based on the available variables. The survey documentation is very useful in operationalizing the variables of interest. The selection of independent variables for secondary data analysis is often based on the underlying theoretical model, existing literature, and availability of variables in the data source. The survey data are rich with respect to demographic, medical, and provider characteristics; however, data related to disease severity, quality of life, and behavioral characteristics may be lacking in national surveys. The diagnostic tests for conducting regression analyses should always be conducted before their implementation. They include evaluating outliers, normality distribution, collinearity, and homoscedasticity (Belsley et al. 2005). Survey regression procedures in SAS, STATA, and SUDAAN can adjust for design complexities in conducting the analyses.

Strengths and weaknesses of national surveys

There are several advantages of using national survey data. The national survey datasets are mostly free or relatively inexpensive. It is easy to obtain national survey data from the NCHS website. The software packages such as SAS and STATA can be effectively used to manage and analyze large national survey data. It takes less time to conduct secondary data analysis than to obtain primary data and perform the analysis. This is true for all studies involving secondary data. The respondents in national surveys are nationally representative and, thus, the findings are generalizable to the population. In addition, most surveys over-sample underrepresented populations (e.g. minorities) and provide adequate sample size for subgroup analyses. The sampling weights can be used to calculate both prevalence and population estimates at the national level. National surveys collect extensive healthcare data on cost, access, and quality, and can be used to address a wide array of research questions at the national level. Variables used for analyses are relatively easy to operationalize due to extensive documentation of data collection and coding procedures. Data-rich national surveys are also useful in adjusting for several characteristics in regression analyses. Surveys such as the NHIS and the MEPS provide an opportunity to analyze data at both person and family levels.

Most of the limitations of national survey are the same as with any secondary data. The variables available for secondary data analysis are developed and defined by data collection agencies. Secondary data analysis is limited to these definitions and variables. Most national surveys are cross-sectional; longitudinal study designs cannot be used with such data. The study based on national survey data is limited to the scope of the original survey. For example, the NAMCS data are limited to visits to office-based physicians in nonfederal settings and are not generalizable to other settings. National surveys may not collect information relevant for some research topics such as those focusing on pharmacist activities. New variables can be created using existing variables; however, there is limited flexibility in redefining the variables. Validity and reliability issues can also pose a problem. For example, self-reported diagnosis data from patients in the MEPS are not the same as the diagnosis data in the NAMCS reported by physicians. Although reliability of most measures may not be an issue, measures based on instruments, such as quality of life, should be evaluated for their reliability in the study sample. National surveys over-sample policy-relevant, underrepresented populations; however, subgroup analyses of other populations may not be possible due to sample size issues. Unlike other secondary data sources, national survey data are released as public use data files years after the data collection. Consequently, survey data may not represent the most recent issues, trends, or practice patterns.

Summary and conclusions

The national surveys provide valuable data to address various cost, access, and quality-of-pharmaceutical-care issues. Selection of the survey data is usually based on the research objectives. Secondary data analysis considerations, such as operationalization of variables, are relatively easy due to the extensive documentation of national surveys. National survey data are based on a multistage complex sampling design. Statistical procedures for secondary data analysis should adjust for survey design complexities. Statistical packages such as SAS and STATA can manage as well as analyze large complex survey data. There are several advantages to using national surveys for research, including cost and time to conduct the research. In addition, the findings from national surveys are generalizable. These strengths have to be weighed against the limitations of national survey data such as variable definitions and sample size issues. With increasing discussion on cost and quality of pharmaceutical care, national surveys provide a cost-effective way to address critical policy issues at the national level.

Review questions/topics

1 Describe national survey data sources that collect data on prescribing patterns.
2 Describe national survey data sources that can examine population-based prevalence patterns.
3 Explain the process involved in calculating sampling errors using the Taylor series linearization and replication methods.
4 Compare and contrast SAS and SUDAAN for analyzing national survey data.
5 What are some strengths and weaknesses of national surveys?

References

Adler GS (1994). A profile of the Medicare Current Beneficiary Survey. *Health Care Financ Rev* 15: 153–63.

Belsley DA, Kuh E, Welsch RE (2005). *Regression Diagnostics*. New York: Wiley.

Botman S, Moore T, Moriarity C, Parsons V (2000). Design and estimation for the National Health Interview Survey, 1995–2004. *Vital Health Stat* 2: 130.

Bryant E, Shimizu I (1988). Sample design, sampling variance, and estimation procedures for the National Ambulatory Medical Care Survey. *Vital Health Stat* 2: 108.

Cochran WG (1977). *Sampling Techniques*. New York: Wiley.

Cohen JW (1997). *Design and Methods of the Medical Expenditure Panel Survey Household Component*. MEPS Methodology Report No. 1. AHRQ Pub No. 97-0026. Rockville, MD: Agency for Healthcare Research and Quality.

Department of Economic and Social Affairs, Statistics Division (2005). *Household Surveys in Developing and Transition Countries*. New York: United Nations Publications.

Jones AL, Dwyer LL, Bercovitz AR, Strahan GW (2009). The National Nursing Home Survey: 2004 overview. *Vital Health Stat* 13: 167.

Kish L (1965). *Survey Sampling*. New York: Wiley.

Korn EL, Graubard BI (1999). *Analysis of Health Surveys*, New York: Wiley.

Kovar M (1989). Data systems of the National Center for Health Statistics. *Vital Health Stat* 1: 23.

Lee ES, Forthofer RN, Lorimor RJ (1989). *Analyzing complex survey data*. Newbury Park, CA: Sage Publications.

McCaig LF, McLemore T (1994). Plan and operation of the National Hospital Ambulatory Medical Survey. Series 1: programs and collection procedures. *Vital Health Stat* 1: 34.

National Center for Health Statistics (1994). Plan and operation of the Third National Health and Nutrition Examination Survey, 1988–94. *Vital Health Stat* 1: 32.

Research Triangle Institute (2005). *SUDAAN 9.0*. Research Triangle Park, NC: Research Triangle Institute.

SAS Institute (2006). *SAS 9.1.3 Procedures Guide*, Vol 4. Cary, NC: SAS Institute.

Sharma R, Chan S, Liu H, Ginsberg C (2001). *Health and health care of the Medicare population: data from the 1997 Medicare Current Beneficiary Survey*. Rockville, MD: Westat.

StataCorp (2007). *Stata Statistical Software: Release 10*. College Station, TX: StataCorp LP.

Thompson KT (2002). *Sampling*. New York: Wiley.

Wolker KM (1985). *Introduction to Variance Estimation*. New York: Springer Verlag.

Online resources

Agency for Healthcare Research and Quality. "Medical Expenditure Panel Survey." Available at: www.meps.ahrq.gov.

Centers for Medicare and Medicaid Services Medicare Current Beneficiary Survey (MCBS). Available at www.cms.hhs.gov/MCBS.

National Center for Health Statistics. "NHANES Web Tutorial." Available at: www.cdc.gov/nchs/tutorials/nhanes/index.htm.

National Center for Health Statistics. "Surveys and Data Collection Systems." Available at www.cdc.gov/nchs/surveys.htm.

Section on Survey Research Methods, American Statistical Association. "Summary of Survey Analysis Software." Available at: www.hcp.med.harvard.edu/statistics/survey-soft.

UCLA Academic Technology Services. "Survey Data Analysis Portal." Available at: http://statcomp.ats.ucla.edu/survey.

15

Program evaluation

Jon C Schommer

Chapter objectives

- To describe the purpose of program evaluation
- To present an approach for conducting effective program evaluations
- To review standards for effective program evaluation
- To address misconceptions regarding the purpose of program evaluation

---

Introduction

Health improvement for individual patients and patient populations is what health professionals including pharmacists strive to achieve. To reach this goal, there is a need to be accountable and committed to achieving measurable health outcomes for the pharmaceutical services and programs. Effective program evaluation is a systematic way to report and improve an organization's actions by involving procedures that are useful, feasible, ethical, and accurate (Centers for Disease Control and Prevention [CDC] 1999).

According to McNamara (2006), program evaluation is helpful to:

- understand, verify, or increase the impact of products or services on customers, clients, or patients
- improve delivery mechanisms to be more efficient and reduce waste
- verify that the organization is implementing programs as originally planned
- facilitate managerial decision-making regarding goals, how to meet goals, and how to determine if goals are being met
- provide information that can be used to verify desired outcomes and for public relations and promoting programs in the community and to sponsors

- conduct valid comparisons between programs to help decide which should be retained or expanded
- fully examine and describe effective programs for replication elsewhere.

The purpose of this chapter is to describe an approach for conducting effective program evaluations, review standards for effective program evaluation, and finally address some of the misconceptions regarding the purpose of program evaluation.

An approach for conducting effective program evaluations

In this section, a step-by-step approach for program evaluation is presented. It should be noted that each step builds upon decisions made at previous steps and that the approach is an iterative process in which adjustments will be made to previous steps before the final process is established.

Step 1: consider goals-based, process-based, and outcomes-based evaluations

The evaluation process begins by engaging stakeholders who would have an investment in what will be learned from an evaluation and who would make decisions based on the information. Three principal groups of stakeholders to consider include: (1) those involved in program operations; (2) those served or affected by the program; and (3) primary users of the evaluation. The scope and level of stakeholder involvement will vary for each program evaluation. Occasionally, stakeholders may use their involvement in an evaluation to distort or bias the evaluation in their favor, so development of trust among stakeholders is essential.

Initially, stakeholders and researchers should address whether the program evaluation would best be designed as a (1) goals-based evaluation, (2) process-based evaluation, or (3) outcomes-based evaluation (McNamara 2006). A goals-based evaluation may be most relevant for the program evaluation of programs established to meet one or more specific goals often described in the original program plans, and if decision makers are most interested in goals-oriented decisions. Such an evaluation would study the extent to which programs are meeting predetermined goals or objectives. An example of a goals-based question would be: "Is pharmacist-provided medication therapy management decreasing hospital readmission for a defined population of patients?"

Alternatively, decision makers might be most interested in making decisions related to how a program works. Process-based evaluations are geared to fully understand how a program produces the results that it does. These evaluations are useful: if programs are long-standing and have changed over time; if employees, patients, or clients report a large number of complaints about the program; or if there appears to be a large number of inefficiencies in delivering program services. Process-based evaluations are also useful for accurately portraying to others how a program truly operates, which is useful for replication of such programs elsewhere. For example, a process-based question would be: "Is a long-standing practice, in which every patient is counseled about a medication at the time of purchase, still providing useful information for patients in light of a new medication therapy management service being offered by the pharmacy?"

Finally, outcomes-based evaluations help understand if an organization is accomplishing the outcomes that are needed by its clients. Outcomes are not just units of services provided or units of products produced. Rather, outcomes are benefits to clients from participation in the program. For example, an outcomes-based question would be: "Is a newly implemented medication therapy management service helping patients manage their medication therapies in terms of what they hold to be important for their health (e.g., access to medicine they need, affordability, and meeting adherence aspirations)?"

Taking time to consider the nature of the decision problem being addressed in terms of being goals based, process based, or outcomes based can help initial discussions with stakeholders be more fruitful, and can lead to consensus about the true nature of the decision problem and resultant research problem to be addressed in the program evaluation.

Step 2: use of theories for guiding evaluations

The next step would be to conduct a systematic literature review to (1) define the domain being addressed, (2) identify previous research conducted in that domain, and (3) identify useful frameworks or theories that could help guide the evaluation. A systematic literature review can not only help to frame the decisions and issues that stakeholders are facing, but also uncover previous work that has already addressed and, in some cases, answered the questions set before the decision makers.

Step 3: formulate problem

With careful and thorough background work, the problem formulation process can be accomplished through a transparent and consensual manner. Part of the process of problem definition includes specifying the objectives of the specific research project or projects that might be undertaken. Each project should have one or more objectives. Subsequent steps should not be taken until each objective can be explicitly stated. The importance of writing down the research problem and specific objectives for the project(s) being undertaken cannot be overstated.

When developing a program evaluation, decision makers often want to know everything about their products, services, or programs. Such an approach can provide findings that turn out to be interesting, but not very actionable. The findings might reduce levels of uncertainty about the health-care program being studied, but provide little understanding of the true problem that is facing the decision makers. Both decision makers who use the findings and the researchers who conduct the program evaluation need to recognize that research does not produce answers or strategies. It produces data that must be interpreted and converted into action plans by the decision makers. Thus, the research must reflect decision makers' priorities and concerns.

It is only when the problem has been clearly defined and the objectives of the evaluation research precisely stated that the evaluation can be designed to generate the information needed in an efficient manner. A detailed under-standing of the overall decision situation should enable researchers, working together with managers and other decision makers, to translate the decision problem into a research problem. A research problem is essentially a restate-ment of the decision problem in research terms.

In developing a research problem, the researcher must make certain that the real decision problem, not just the symptoms, is being addressed. There have been many cases in which poor problem definition led to poor research problem definition with unfortunate consequences (Churchill 1995). Con-sequently, it is important for the decision maker and the researcher to have a meeting at which the decision maker describes the problem and the informa-tion that is needed. The researcher then drafts a statement describing his or her understanding of the problem. The statement should include, but is not lim-ited to, the following items:

- *Action*: the actions that are contemplated on the basis of the research.
- *Origin*: the events that led to a need for the decision to act; even though the events may not directly affect the research that is conducted, they help the researcher understand more deeply the nature of the research problem.
- *Information*: the questions that the decision maker needs to have answered in order to take one of the contemplated courses of action.
- *Use*: a section that explains how each piece of information will be used to help make the action decision; supplying logical reasons for each piece of the research ensures that the questions make sense in light of the action to be taken.
- *Targets and subgroups*: a section that describes from whom the information must be gathered. Specifying the target groups helps the researcher design an appropriate sample for the research project.

- *Logistics*: a section that gives approximate estimates of the time and money that are available to conduct the research; both of these factors will affect the techniques finally chosen.

This written statement should be submitted to the decision maker for his or her approval, including signature and date. This step will help assure agreement on the purpose of the evaluation research before the research is designed.

Getting to this point in a program evaluation is not easy. Decision makers who will be using information from the evaluation and researchers who will be conducting the evaluation often have different training, priorities, and viewpoints. To facilitate communication about identification of the decision problem and translating it into a research problem for a program evaluation, spending time with stakeholders to describe the program being evaluated can be useful.

Step 4: determine research design

Research design depends on how much is already known about the problem. If previous steps reveal that relatively little is known about the phenomenon to be investigated, exploratory research is warranted (Churchill 1995). Exploratory research may involve review of existing data, interviewing knowledgeable people, conducting focus groups, conducting in-depth interviews, or investigating literature that discusses similar cases. One of the benefits of exploratory research is its flexibility. It allows the researcher to follow the leads that may develop during the process. If the problem is precisely and unambiguously formulated, then descriptive or causal (experimental or quasi-experimental) research is better suited for program evaluation. In these research designs, data collection is rigidly specified, with respect to both data collection forms and the sample design. The advantage of these methods is that they allow testing hypotheses, making population estimations, and testing cause–effect relationships.

Step 5: design data collection and forms

Quite often, the information needed for a program evaluation can be obtained from an organization's own databases or internal records, or in published documents such as government reports or publicly available databases. However, some evaluations must depend on primary data, which are collected specifically for the study. At this point, a number of decisions must be made for the evaluation research. For example, should the data be collected by observation or questionnaire? Should the form be structured as a fixed set of answers or should they be open-ended? Should the purpose of the study be made clear to the respondent, or should the study objectives be disguised? Each decision will depend upon the research problem and study objectives as well as the time and resources available for conducting the study.

When theories, models, and frameworks are used to help explain phenomena, constructs are used as the building blocks. Conceptual definitions provide information about how a given construct is defined in terms of other constructs in the theory, model, or framework. An operational definition describes how the construct is to be measured. To help assure that results among studies are comparable, operational definitions should be theory driven and based on the relevant construct domain. A theory base will help interpret the results with other studies and will allow for comparisons with other programs with possible longitudinal evaluation over time.

Step 6: design sample and collect data

In designing the study sample, the researcher must specify the sampling frame, the sample selection process, and the size of the sample. For some program evaluations, it would be desirable to study outcomes over a relatively long period of time in order to answer relevant questions (longitudinal effects). To conduct such research, panels can be used. A panel is a fixed sample of elements and relies on repeated measurements of the same variables. Although the major advantage of a panel is analytical, panels also have some advantages in terms of the types of information collected and the accuracy of information collected. However, panels are costly in time, money, and the attrition of sample members, and often are not representative of the population of interest.

Step 7: analyze and interpret data

Researchers may amass a mountain of data, but these data are useless unless the findings are analyzed and the results interpreted in the light of the problem at hand (Churchill 1995; CDC 1999; US Department of Health and Human Services [HHS] 2005). Data analysis involves multiple steps. First, data collection forms must be reviewed to ensure that they are complete and consistent with instructions provided in the forms. Editing of forms may be needed to correct errors, convert responses into the units specified in the instructions, and address item non-responses. The next step after editing is coding which involves assigning numbers (or other interpretable values such as text) to each of the answers, so that they may be analyzed, typically by computer. The final step in analyzing data is tabulation; it refers to the orderly arrangement of data in a table or other summary format by counting the frequency of responses to each question and by conducting statistical analysis for testing hypothesized differences or relationships. The statistical tests applied to the data, if any, are somewhat unique to the particular sampling procedures and data collection instruments used in the research. Statistical tests should be anticipated before data collection begins so that the data and analyses will be germane to the problem as specified for the study. It should be noted that data analysis in program evaluation can pose challenges. Some of the common challenges include (1) cluster bias, (2) nonlinear relationships, and (3) case mix.

Cluster bias

Whenever physician prescribing is studied using the patient as the unit of analysis, cluster bias could become problematic for analysis. For example, if one physician in the study prescribed medications for a disproportionate number of patients in relation to another physician in the study, results could be biased by this difference. As the observations are correlated, standard errors of estimated coefficients are biased downward, resulting in an increased likelihood of significant coefficients. Adjusting for cluster bias increases standard errors by accounting for the correlation between observations (Bieler and Williams 1997). In addition, program evaluation data may have a nested structure, e.g., repeated observations nested among individuals or individuals nested within groups such as health plans or geographic regions. The use of hierarchical linear models can be used to help improve the estimation of individual effects in these data structures (Bryk and Raudenbush 1992).

Nonlinear relationships

Relationships among study variables are not always linear and may be hard to detect. For example, individuals might view too much or too little of a particular healthcare service as having low value or utility. Service provision in the right amount would be viewed as having the highest value or utility. Such an inverted U-shaped relationship can be modeled by studying both the main effect and the squared term for the value or utility score. For an inverted U-shaped curve, the main effect of the service provision would exhibit a positive relationship with the value or utility score and the squared term would show a negative relationship. A good way to identify nonlinear relationships is to prepare scatter plots for pairs of variables under study and to view the shape of each plot. Examples of other nonlinear relationships include threshold effects and exponential effects that might be found in the data.

Case mix

Finally, when comparing results among groups of individuals who belong to different groups or categories, it is important to control for case mix. Case mix refers to the different study participant attributes that might be apparent in different groups or categories (Johnson 1997). To the extent that patients self-select or are assigned to different comparison groups in nonrandom ways, comparisons between the groups may be biased. Case-mix control variables such as age, gender, socioeconomic status, disease severity, or physical functioning should also be considered for inclusion in the study. Methods used to control or adjust for case-mix variables can then be used (e.g., restriction, stratification, matching, statistical adjustment) (Lobo 2003).

Step 8: prepare the research report

Interpretations of findings into actionable conclusions are justified when they are linked to the evidence gathered and judged against standards set by the stakeholders. Stakeholders must agree that conclusions are justified before they use the evaluation results with confidence. A research report is a useful communication tool for sharing findings with stakeholders, for obtaining feedback on draft versions, and for developing a final version that can be agreed upon by stakeholders. In some cases, alternate views and interpretations may be included in the report to help maintain transparency and fair balance. The research report is all that most stakeholders will see of the program evaluation, and it becomes the standard by which the research is judged. A general outline for research reports consists of (1) title page, (2) table of contents, (3) executive summary, (4) introduction, (5) methods, (6) results, (7) discussion of results within the context of the purpose of the study, (8) conclusions and recommendation, and (9) appendices (e.g., copies of data collection forms, detailed calculations used for analysis, tables not included in the main part of the report, bibliography).

Standards for effective program evaluation

So far, this chapter has described the purpose of program evaluation and proposed an approach for conducting effective program evaluations. However, for program evaluation research to become a practical tool that stakeholders and decision makers can use to inform programs' efforts and assess their impact, these evaluations need to be integrated into day-to-day planning, implementation, and management for programs (CDC 1999; HHS 2005).

High-quality program evaluations typically use science-based research as a basis for decision-making and action, but also focus on expanding the quest for social equity in program access, performance that is efficient and effective, and achieving outcomes that benefit society. The following standards are proposed for making sound and fair evaluations practical: (1) utility, (2) feasibility, (3) propriety, and (4) accuracy. The standards provide guidelines to follow when deciding among evaluation options and help avoid creating imbalanced action plans (e.g., pursuit of a plan that is accurate and feasible but not useful, or one that is useful and accurate but is not feasible). The standards are guiding principles, not mechanical rules. Thus, whether a given standard has been addressed adequately in a particular situation is a matter of judgment (CDC 1999; HHS 2005).

Utility

Utility standards ensure that information needs of evaluation users are satisfied. There are seven specific utility standards (CDC 1999; HHS 2005):

1   *Stakeholder identification*: individuals involved in or affected by the evaluation should be identified so that their needs can be addressed.
2   *Evaluator credibility*: individuals conducting the evaluation should be trustworthy and competent in performing the evaluation.
3   *Information scope and selection*: information collected should address pertinent questions regarding the program, and be responsive to the needs and interests of clients and other specified stakeholders.
4   *Values identification*: perspectives, procedures, and rationale used to interpret findings should be described.
5   *Report clarity*: reports should clearly describe the program being evaluated, including its context and purposes, procedures, and findings of the evaluation.
6   *Report timeliness and dissemination*: substantial interim findings and evaluation reports should be disseminated to intended users.
7   *Evaluation impact*: evaluations should be planned, conducted, and reported in ways that encourage follow-through by stakeholders.

Feasibility

Feasibility standards ensure that the evaluation is viable and pragmatic. There are three specific feasibility standards (CDC 1999; HHS 2005):

1   *Practical procedures*: evaluation procedures should be practical while needed information is being obtained to keep program disruption to a minimum.
2   *Political viability*: consideration should be given to the varied positions of interest groups so that their cooperation can be obtained and maintained.
3   *Cost-effectiveness*: the evaluation should be efficient and provide valuable information to justify expended resources.

Propriety

Propriety standards ensure that the evaluation is conducted with regard to the rights and interests of those involved and affected. There are eight propriety standards (CDC 1999; HHS 2005):

1 *Service orientation*: the evaluation should be designed to assist organizations in addressing and serving the needs of the target participants.
2 *Formal agreements*: principal parties for the evaluation should agree in writing to their obligations.
3 *Rights of humans*: the rights and welfare of human participants must be respected and protected.
4 *Human interactions*: interactions with other individuals associated with an evaluation should be respectful so that participants are not threatened or harmed.
5 *Complete and fair assessment*: the evaluation should record strengths and weaknesses of the program completely and fairly so that strengths can be enhanced and problem areas addressed.
6 *Disclosure of findings*: full evaluation findings with pertinent limitations should be made accessible to people affected by the evaluation and any others with expressed legal rights to receive the results.
7 *Conflict of interest*: should be handled openly and honestly.
8 *Fiscal responsibility*: allocation and expenditure of resources should reflect sound accountability procedures.

Accuracy

Accuracy standards ensure that the evaluation produces findings that are considered correct. There are 12 accuracy standards (CDC 1999; HHS 2005):

1 *Program documentation*: the program being evaluated should be described clearly and accurately.
2 *Context analysis*: the context in which the program exists should be examined in enough detail to identify probable influences on the program.
3 *Purposes and procedures*: purposes and procedures for the documentation should be described in sufficient detail so that they can be monitored and audited.
4 *Defensible information sources*: sources of information should be described in enough detail to assess the adequacy of the data.
5 *Valid information*: information-gathering procedures should be described to ensure valid interpretations.

6   *Reliable information*: information-gathering procedures should be described to ensure sufficiently reliable information for intended use.
7   *Systematic information*: information collected, processed, and reported should be systematically reviewed and any errors corrected.
8   *Analysis of quantitative information*: quantitative analysis techniques should be described so that others can replicate the work.
9   *Analysis of qualitative information*: qualitative analysis techniques should be described so that others can replicate the work.
10 *Justified conclusions*: conclusions should be explicitly justified for stakeholders' assessment.
11 *Impartial reporting*: reporting procedures should guard against distortion caused by personal feelings and biases of any party.
12 *Meta-evaluation*: the evaluation should be formatively and summatively assessed against these and other pertinent standards in order to, upon completion, enable close examination of its strengths and weaknesses by stakeholders.

Misconceptions regarding the purpose of program evaluation

To bring this chapter to a close, a few misconceptions about the purpose of program evaluations are highlighted (McNamara 2006). These misconceptions represent the viewpoints of some stakeholders as program evaluation projects are considered. An understanding of these misconceptions can help researchers as they plan communication strategies for initiating a program evaluation and for leading the process of problem formulation at the start of projects.

Some people believe evaluation is a useless activity that generates a lot of boring data with useless conclusions (McNamara 2006)

If a stakeholder's past experience with program evaluations was one in which previous methods were chosen largely on the basis of maximizing scientific accuracy, reliability, and validity, he or she may hold this view. That approach can generate extensive data from which very carefully chosen conclusions were drawn, application was avoided, and, as a result, evaluation reports tended to state the obvious and decision makers were left disappointed and skeptical about the value of the evaluation. To help change these individuals' opinions, one may need to explicitly show how the planned program evaluation will provide utility, relevance, and practicality, as well as scientific validity, to the final report.

Some people believe that evaluation is about proving the success or failure of a program (McNamara 2006)

This misconception assumes that success lies in implementing the perfect program and never having to hear from employees, customers, clients, or patients again about how to improve it or change it. This does not happen in real life. Success is remaining open to continuing feedback and adjusting programs accordingly. Evaluation gives this feedback. To help avoid this misconception in program evaluation, special care should be devoted to how the research problem is stated and written. It should focus on gaining feedback that can be applied for program improvement and specific decisions being faced.

Some people believe that evaluation is a highly unique and complex process that occurs at a certain time in a certain way, and almost always includes the use of outside experts (McNamara 2006)

Some people believe that, in order to be able to conduct a program evaluation and use the findings, one must have expertise in such research. If this is the case, the researcher can help outline how scientific findings will be translated so that stakeholders can use the findings for decision-making and application. The commitment from the stakeholders who will be using the findings is to be actively engaged in the problem formulation process and to make judgments about the usefulness of the findings as they review drafts of the final report. Such engagement in the process will help researchers conduct the program evaluation design, implement, and report findings from the evaluation in a way that is understandable, verifiable, and actionable.

Summary and conclusions

The purpose of this chapter was to: (1) describe the purpose of program evaluation; (2) present an approach for conducting effective program evaluations; (3) review standards for effective program evaluation; and (4) address misconceptions about the purpose of program evaluation. In summary, effective program evaluation is a systematic way to report and improve an organization's actions by involving procedures that are useful, feasible, ethical, and accurate. A multiple-step, iterative approach for program evaluation includes: (1) consideration of goals-based, process-based, and outcomes-based evaluations; (2) use of theories for guiding evaluations; (3) problem formulation; (4) determination of research design; (5) design of data collection and forms; (6) sample design and data collection; (7) analysis and interpretation of data; and (8) preparation of the research report. This chapter also proposed four standards for effective program evaluation: (1) utility, (2) feasibility, (3) propriety, and (4) accuracy. Effective program evaluation is a systematic way to report and improve an organization's actions by involving transparent procedures that help make judgments about the program, improve program effectiveness, and aid in future program development.

Review questions

1 What is the purpose of program evaluation?

2 What is the "research request step" and why is it used?

3 What are the eight steps to the program evaluation approach outlined in this chapter?

4 Why is the "problem formulation" step so important for program evaluation research?

5 What are four standards for effective program evaluation?

6 How can program evaluation be useful for making decisions in pharmacy practice?

References

Bieler GS, Williams RL (1997). *Application of the SUDAAN Software Package to Clustered Data Problems in Pharmaceutical Research*. Research Triangle Park, NC: Research Triangle Institute.

Bryk AS, Raudenbush SW (1992). *Hierarchical Linear Models*. Advanced Quantitative Techniques in the Social Sciences Series, Vol 1. Newbury Park, CA: Sage Publications.

Centers for Disease Control and Prevention (CDC) (1999). Framework for program evaluation in public health. Morbid Mortal Weekly Rep 48(RR11): 1–40.

Churchill GA (1995). *Marketing Research, Methodological Foundations*, 6th edn. New York: The Dryden Press.

Johnson JA (1997). Patient satisfaction. In: *Pharmacoeconomics and Outcomes: Applications for patient care*. Myrtle Beach, SC: American College of Clinical Pharmacy, 111–54.

Lobo FS (2003). Assessment of channeling bias and its impact on interpretation of outcomes in observational studies: the case of the cyclooxygenase-2 (COX-2) inhibitors. Unpublished doctoral dissertation, University of Minnesota.

McNamara C (2006). *Field Guide to Nonprofit Program Design, Marketing and Evaluation*, 4th edn. Minneapolis: Authenticity Consulting.

US Department of Health and Human Services (HHS) (2005). *Introduction to Program Evaluation for Public Health Programs: A self-study guide*. Office of the Director, Office of Strategy and Innovation. Atlanta, GA: Centers for Disease Control and Prevention.

Online resources

Framework for Program Evaluation. Available at: www.cdc.gov/eval/framework.htm.

Framework for Program Evaluation in Public Health. Available at: www.cdc.gov/mmwr/preview/mmwrhtml/rr4811a1.htm.

Knowing What Works in Healthcare: A roadmap for the nation. Available at: www.nap.edu/catalog/12038.html.

16

The future of pharmaceutical policy research

Albert I Wertheimer

Chapter objectives

- To present the importance of pharmaceutical policy research
- To discuss current methods and metrics in pharmaceutical policy research
- To discuss current issues and future direction of pharmaceutical policy and practice research

Introduction

It has been more than 40 years since pharmaceutical issues have been in the public policy arena, with the inclusion of a pharmaceutical benefit in the Medicaid program established in 1965. This is when the researchers and policy makers first became interested in understanding how pharmaceutical issues are dealt with in the health policy arena. Since that time, there has been growth and development of health maintenance organizations (HMOs), pharmacy benefit management companies (PBMs), and more recently the inclusion of a drug benefit in the US Medicare program as well as the emergence of managed care organizations (MCOs). This chapter examines some of the major issues, trends, and forces that will determine the future of pharmaceutical policy.

Future direction for pharmaceutical policy research

The pharmaceutical realm can no longer be looked at in isolation because it is an integral component of the overall healthcare delivery system. The bottom-line impact in health policy is that any change or activity in any sector of the health field can and usually does affect all of the other aspects. For example, a change in hospital length of stay characteristics to a shorter average time means that a sicker patient will be going home and drug therapy previously used only in the in-patient setting may now be used in the patient's home. Another example is how behaviors may be manipulated through a series of incentives and disincentives for the prescribing and use of the most cost-effective drug products. There are many other examples, but it is safe to say that an understanding of the fit of the pharmaceutical sector into the overall healthcare delivery system can be of great value in evaluating the current structure and in planning future healthcare delivery system designs (Winkler 2007).

At one time, pharmaceuticals were of interest and importance only to physicians, patients, and the pharmaceutical industry. During the last 40 years, they have become involved in labor union collective bargaining agreements, retirement planning, and major governmental deliberations regarding the cost and nature of several federal entitlement programs. At the same time, an originally very small generic pharmaceutical industry became a giant and now represents more than half of the prescriptions dispensed in the USA. As one might imagine, this automatically leads to other sectors being involved in policy matters about pharmaceuticals such as the patent office and a number of trade organizations (Seget 2008).

Despite all the headlines and news reports about various happenings and events with pharmaceuticals, the industry is still a very small one and this should be kept in perspective. The total sales in agriculture, defense, automobiles, construction, chemicals, clothing, and many other industries are far larger than pharmaceutical sales. Today, that arena is changing greatly with increasing emphasis in the area of biotechnology at the expense of an almost total focus on synthetic chemically prepared smaller molecules in the search for new and valuable therapeutic agents (Cockburn 2004).

The importance of pharmaceutical policy research may be best described by stating that pharmaceutical spending has seen a rise from 8.7 percent in 1982 to 12.4 percent in 2005 of the total US annual expenditures in the healthcare field (Civan and Köksal 2010). Needless to say, that is not a small number and, although it is tiny compared with the amount of money spent on hospitals and physician services, it is a huge number where efficiencies and wise policy decisions can yield savings in the expenditures for pharmaceuticals and in those other spheres in the global healthcare arena (Ess et al. 2003).

Methods and metrics

There is a need for data (and lots of them) to be able to see the present situation before we are able to determine whether something can be changed for the good. This brings up an important point: all too often health policies are changed by new government administrations or new directors, or after some catastrophe or scandal. Change for the sake of change is to be avoided. It might be an acceptable approach for politicians to demonstrate to the electorate that something is being done, but healthcare professionals and researchers must avoid and discourage that strategy. That old adage that you don't fix something unless it is broken is operational here as much as in any other sector.

Data sources

Healthcare data are obtained from a large variety of sources. A very common method of obtaining data is through the employment of prospective studies, where experimental designs are used or observational data are collected as events and outcomes occur. This may be obtained from medical records, patient interviews, or questionnaires. The advantages of such a study are that it is designed by the investigator and therefore collects all the fields of information that are needed, and moreover the level of accuracy and consistency of the data are assured. The negative features are the obvious ones: an enterprise such as this could take a great deal of time and is most likely quite costly. If either resources or time, or both, does not permit the use of a prospective study, then the researchers turn to a retrospective study. This is use of data that have already been accumulated or collected, usually for some other purpose. The positive side of this is that massive quantities of data may be available almost immediately and usually at little or no cost. The downside is that there may be issues with the standards for precision and accuracy of the original study or of the goals and biases of the investigators. Moreover the researchers must accept the data as they are, which means that some important and critical variables may be missing but investigators feel that having most of the needed data is better than none.

Often, the data that are found are a series of unrelated studies conducted in different locales with different populations and often with slightly different methodologies. Fortunately for us, there are statistical methods to combine those numerous smaller studies and to synthetically create a larger panel or sample size that may be used for policy analysis. This synthetic combination is called meta-analysis. It is described in great detail in statistics and research methods textbooks (Gold et al. 1996).

Generally, retrospective data are available in one or two categories. Administrative data are usually the easiest to obtain and most usable. A more detailed retrospective data source would be from medical records themselves. Where medical records have been converted to electronic form, the task is a bit easier because the charts are then laid out in a uniform manner (e.g., laboratory findings are always found in the same place). Handwritten medical records generally require the use of medical records personnel, nurses, or others to abstract the charts by going from page to page manually. Needless to say, this is costly, time-consuming, and rather inefficient. But perhaps worst of all is the fact that many medical records are far from complete.

In recent times, health services researchers have turned to large databases for needed information. MCOs often enroll more than a million members and, naturally, these would be sources to which one would turn in the quest for retrospective data. Some MCOs have been more progressive and are known to have outstanding computerized records, and those would be the first places to turn. Some of these include the Kaiser–Permanente Health Plans, Group Health of Puget Sound in Seattle, Washington, the Henry Ford Health System in Detroit, and a number of others. Another enormous source of data is that of the Veterans' Administration (VA) health system which has been computerized for a number of years. Naturally, the VA patient population is not representative of the US population as a whole, because it contains an overwhelming proportion of aging males, but it should not be ignored since the VA system is research friendly and large. This database has almost all medical records for its veteran population and, as the patients receive free or nearly free care, there is very little leakage from the system.

The Medicare program makes available several databases for the use of researchers. As there are more than 40 million senior citizens enrolled in the Medicare program in addition to small numbers of disabled people, as well as patients with end-stage kidney disease and a few other medical conditions affecting those below the age of 65, it is the most desirable database source. There are numerous databases to use either singly or linked with other databases (Chan et al. 2001).

A few other notable sources include:

- *Hospital Compare*: quality-of-care data for selected medical and surgical procedures for Medicare beneficiaries
- *Nursing Home Compare*: which is similar to its hospital counterpart but only for the nursing home population
- *Medigap data*: information for patients with Medicare and additional private health insurance
- *Medicare Advantage*: a managed care-type program for Medicare beneficiaries with comprehensive data
- *Medicare Current Beneficiary Survey*: a complete listing of demographic and health characteristics of the Medicare population (see Chapter 11 for details)
- *SEER (Surveillance, Epidemiology and End Result) database*: comprehensive morbidity and treatment information for the entire Medicare population. This is probably the gold prize of the Medicare program for researchers (Riley 2009; Yabroff et al. 2009).

The Medicaid program is quite different because different states have varying eligibility requirements and program characteristics. In addition, researchers are obliged to obtain state Medicaid database information on an individual basis. It is possible to link and combine state Medicaid database information, but the Medicaid program has a number of major shortcomings, one of the most important being that some people remain in the program for months, drop out, and then re-enter later so that the records may be incomplete. Nevertheless, it is a valuable source of information about the treatment, cost, and morbidity patterns of medically indigent individuals in the USA.

Other data sources include the American Hospital Association hospital discharge information, as well as proprietary databases offered by a number of suppliers covering most specialty areas. The most comprehensive information about pharmaceuticals, including costs, prescribing patterns, trends, market share, and indications, among many other variables, comes from IMS Health of Plymouth Meeting, Pennsylvania and other commercial sources (see Chapter 12 for additional details).

The three goals of studies of such databases are to reduce costs, increase efficiency, and increase the quality of care of medical and health services in the USA. This is also done through the use of standardized measures from a number of organizations. The previously mentioned Medicare and Medicaid programs have participation requirements for providers in institutions that must be met before they are eligible to accept program-sponsored beneficiaries and to be paid for services provided to them. In addition, there is an organization that accredits hospitals, called the "Joint Commission," which provides accreditation of hospitals and healthcare organizations, generally referred to as JCAHO (pronounced JAY–KO). It reviews hospital performance on a number of variables and actually visits or inspects hospitals every few years. MCOs are evaluated by the National Committee for Quality Assurance (NCQA) through its Healthcare Effectiveness Data and Information Set (HEDIS) measures. Similarly, MCOs have lengthy questionnaires to complete and self-studies to perform, and they are the subject of periodic inspections (Scanlon and Hendrix 1998; Viswanathan and Salmon 2006).

Measuring quality

The earliest quality studies were pioneered by Ernest Codman, using his end-results approach (Neuhauser 2002) and, later, Donabedian provided the framework for quality measures. Donabedian (1996) suggested three levels of evaluation. The first and simplest is a straightforward structure measure where, for example, one can see whether or not there are fire extinguishers in nursing homes. It is rather obvious that having the fire extinguishers present is not a guarantee that an effective job will be done to counter any fires or to prevent them. Therefore, a second-level of evaluation is in order and that is a process measure. Here, one evaluates whether a nursing home with fire extinguishers has trained some of their personnel to use them and has a fire emergency plan with established responsibilities and practice drills. Although this provides more comfort and confidence, it still does not answer the ultimate question of how well that nursing home does if a fire breaks out. That highest level of evaluation is referred to as an outcome evaluation and would require measuring the results when an actual fire had taken place. Had the staff been able to safely remove all the patients and extinguish the fire with a minimum of damage or trauma? Outcome measures are not always possible and, when they are, they require time and money, so it is often necessary to depend on a structure or process consideration for decision-making (Donabedian 1996).

In addition to the accreditation of facilities, it is also necessary to evaluate competence of practitioners. Previously this was not attempted at all but, during the last 20 years, most health professions have required mandatory continuing education participation. It has been assumed, probably incorrectly, that if a physician or pharmacist or other healthcare provider were to sit through 15 hours of coursework each year he or she would be an up-to-date practitioner. It had been shown that this is not the case and in fact this has been referred to as not continuing and not education. Several medical specialties have gone further and require periodic examinations to maintain specialty status of providers. Boards of pharmacy in the 50 states and the District of Columbia administer licensure examinations and process the licensure and documents for continuing education responsibility completion. Critics of this, our current situation, complained that this does not reassure the public or guarantee continuing competency, and the only way to do that is through periodic licensure examinations. Therefore, it is not too unlikely to expect that periodic licensure examinations will be required some time in the future.

Policy issues

During the twentieth century there have been a number of important policy decisions and policy battles that have taken place. Perhaps there is a need to learn something from looking at several of these.

Through the 1950s medicines were sold exclusively in pharmacies. They were not found in convenience stores, grocery stores, vending machines, or other retail establishments. However, a food chain began selling aspirin and some other over-the-counter medicines, and was challenged by pharmacy organizations. The result of this battle was an embarrassment to the pharmacy establishment because the courts declared that the over-the-counter products were sufficiently safe and that their labeling enabled safe use without any professional involvement. This immediately opened the door for a massive number of nonpharmacy retail outlets to begin selling a wide variety of over-the-counter drug products.

Another major battle happened around the same time when the chain stores began growing and sought pharmacy licenses in numerous states. The state boards of pharmacy, made up mostly of independent pharmacy owners, resisted this "invasion" by the chains and fierce battles were fought in many states. Today, one only needs to look at any street corner to see the result of that confrontation; in fact, the chains continue to grow, often at the expense of the market share of smaller independent pharmacies that seem to be closing. The argument was that chain stores were not owned by pharmacists and therefore there was no one to uphold the important professional standards that are supposedly found in independent pharmacies. The result of this confrontation was that every pharmacy, independent of its ownership, must appoint one pharmacist as the person responsible for all professional activity within the premises. That pharmacist can be the owner or an employee.

The regulations in the 50 states required that pharmacists dispense a drug exactly as the prescription was written, so that, if a specific brand name is indicated, the pharmacist would be obliged to dispense that specific drug product. At this time, in the early 1970s, at the start of the rise of the generic industry, it became possible for the pharmacist to be in a position to save money for the patient by performing generic substitution and dispensing a less costly product to the patient. Naturally, this was resisted by the branded manufacturers and battles broke out in nearly all states. In the end, the restrictive practices constraining the pharmacist were removed and today the pharmacist is able to dispense a generic product unless the physician indicates that a specific brand is medically necessary. Most MCOs and health insurers will pay for generic drugs only when those versions become available at the time of the patent expiration. In such cases, the pharmacist is required by his or her contract with the health insurer or MCO to dispense the generic drug to the patient unless the patient is willing to make a substantially larger copayment (Smith 1991).

Unresolved policy issues

Evidence-based practice

There are a number of unsolved policy questions facing pharmacy and the current healthcare system. One of the more important ones is evidence-based practice, a growing body of knowledge that certain practices lead to optimal outcomes. There has been no standardized means accepted to induce physicians and/or pharmacists and others to comply with these evidence-based practices, often called best practices. Some argue that so-called best practice is not optimal for all patient categories and that the physician or caregiver needs greater latitude to decide what is best for each individual patient. This will probably be settled through the financial realm in the not-too-distant future where the health insurers or MCOs will dictate certain practice characteristics and policies that the clinicians will be required to follow in order to be paid.

Comparative effectiveness research

Another important unresolved policy issue is comparative effectiveness research. This extension of health technology assessment has captured the attention of most stakeholders today in 2009–10. Until the present, new drugs are tested against placebos or inert products and, if the new drug demonstrates safety and effectiveness, it is approved for marketing. However, there is no evidence of whether that drug is equal to, superior, or inferior to the other several drugs in the same therapeutic category already on the market. Comparative effectiveness research is also relevant for surgical procedures and use of diagnostic and test items and for, essentially, virtually all interventions. How, then, can a prescriber determine what is the best product to use for his or her patients? Many are worried and all companies and organizations are concerned. The one constituency that has a great deal at stake is the group of patients with complex, multiple medical issues that may require individualized or personalized medical treatment. These are not the average patients for which disease management protocols or treatment algorithms were designed. At the moment, it is too soon to indicate where this unresolved issue will end and what the results will look like (Institute of Medicine 2009).

Pharmacist practice limits

This is a huge unresolved matter. There are a number of policy issues and questions in this area. For example, pharmacists have argued that they are available 24/7 and may be contacted without prior appointment, and it would seem only reasonable that they should be permitted to prescribe some limited number of drugs for acute situations and that they should be enabled to administer vaccinations and other immunizations to patients. The argument is that: the use of the pharmacist for these services is less costly than using a physician; this enables the physician time to deal with more serious and complex cases; and the pharmacist is available after-hours and weekends, times when many physicians' practices are closed.

Mail service pharmacy

The VA started mailing prescription drug refills to patients in the 1960s to save them the time and expense of having to come to a VA medical centre, possibly in another community, and waiting sometimes many hours for the prescription to be ready. Others in the private sector seized on this opportunity and the mail service pharmacy industry was born and again encountered fierce opposition by community pharmacists, and this time also by the chain pharmacies. Today mail service pharmacies appear to be a growth industry and they are owned and operated primarily by pharmacy PBMs and by chain pharmacy organizations.

Retail pharmacists argue that the mail service format does not offer the one-to-one interaction and personal counseling obtainable only in face-to-face encounters at the community pharmacy. The mail service camp counters that their level of automation and controls achieves an error rate far superior to that found in any community pharmacy and that their 24/7 800 telephone number service provides a solution to any patient dilemma or problem. At the moment it appears that the mail service pharmacies have demonstrated a cost-effective rationale for their existence with possible further growth in this sector.

Medicare Part D

In 2003, the Medicare Part D program was legislated. The full program went into effect in 2006 and patients receive assistance until they reach about US$2200 in expenditures. At that point, they are in what is generally referred to as the doughnut hole and individual Medicare beneficiaries are required to pay 100 percent out of pocket for their prescriptions until they reach about US$5300 in expenditure. The federal government program pays the vast majority of all further prescription medicine costs. This program design requires out-of-pocket expenditures for the first deductible portion, followed by the 25 percent copayment level, and then to about US$3000 in the doughnut hole. This is a great deal of money for many Medicare beneficiaries who are unable to pay for their medicines within this scheme.

Numerous proposals have been offered to eliminate or reduce the financial burden of the doughnut hole, to require the use of generic drugs where available, and to permit direct price negotiation and purchasing of the drugs by the federal government, which would enable them to use their enormous bargaining power to reduce costs. It is likely that changes in the existing program from the Republican administration will be changed by the current Democratic administration in Washington. It is too difficult and complex to speculate on the final configuration of an altered Medicare Part D program (Kilian and Stubbings 2007).

Pharmacy practice

A host of policy issues in the pharmacy practice arena remains. Some of these include the use of automation and robotics in licensed pharmacies to assist in making the prescription dispensing function a lower cost activity. There are several other related policy matters. One of these involves the use of para-professionals, also called pharmacy technicians. The chains and other owners of pharmacies want to be able to use a greater ratio of technicians to phar-macists so that the work can be done with the employment of fewer, costly, registered pharmacists. State board regulations limit the supervision of technicians to only a few by any pharmacist in order to protect the public health. Usually changes are made when someone demonstrates that a practice can be performed safely and economically, and yet such a demonstration would violate regulations or laws today. Therefore probably political prowess will determine the results here. As chain pharmacies are considered to be or are capable of being large donors to political election and re-election cam-paigns, it would be a good bet that the current regulations will be eased.

Another front in that same battle revolves around the regulation that a licensed, registered pharmacist must be present during all opening hours of a pharmacy. Store owners would very much like to operate while the pharma-cist takes a break or has an uninterrupted meal. It is not a great leap of logic to surmise that this innocent practice can lead to a request to boards of pharmacy to permit the dispensing of prescriptions previously checked by the pharma-cist, even if the pharmacist is not present at the time of dispensing to the patient. The bottom line would be that the owners would be relieved if the pharmacist were not required to be present during all hours that the pharmacy is open for business.

A concept that continues to be brought up often in deliberations on this topic is that of the "central fill" concept. This is similar to the old method where one brought exposed photographic film to the pharmacy or supermar-ket and it was picked up by a courier and taken to a central laboratory, where the film was developed and printed and returned to the pharmacy or super-market for the customer to retrieve it. Some have suggested that many unprof-itable locations today could serve as depots for patients to bring in the prescription, and then the prescription would be dispensed at a central fill location and returned to the original pharmacy to be distributed to the patient within 1 or 2 hours. The design would look like a hub with many spokes and that centre hub would be the central fill facility. Perhaps one central fill facility could supply a dozen or more pharmacies located in one section or area of the city. Although the concept is being implemented on a smaller scale, economic and financial impact can be determined to implement such concepts on a wider scale. This would make a very compelling argument to governments and other payers for pharmaceutical services and products that the central fill concept has the potential to significantly reduce their expenditures for prescription drugs.

Another persistent policy question is that of a third class of drugs. Pharmacy has long sought control over such a category of products that would be unavailable from other retail outlet types. The US Food and Drug Administration (FDA) has repeatedly argued that prescription legend drugs require a physician's prescription and should be carefully distributed by pharmacies, but over-the-counter drugs have proven safety records over many years of use, along with quite comprehensive consumer-understandable labeling and the presence of an 800 telephone number for further questions or information. Given this background, the FDA does not support a third category of drugs. However, pharmacists hope that drugs newly moved to the over-the-counter (OTC) category from their previous prescription-only status might be part of that third category for a limited period of time, until their safe and effective use in the OTC category can be established and until such time that they would be moved into the full OTC category. This has been a debated topic for more than a quarter of a century and it does not look like it is any closer to resolution at the moment. It should be pointed out that many other developed countries have three or four categories of pharmaceutical products similar to the situation strived for by US pharmacists.

Money

Paying the pharmacist has always been a spirited topic. Pharmacy moved away from a mark-up system of pricing in the 1950s and 1960s toward the American Pharmacist Association's (APhA's) professional dispensing fee method, which is the principal system of contemporary payment. Now some parties are floating the idea of a pay-for-performance payment system. This is tied to the flawed "pharmaceutical care" concept, where the pharmacist was to be responsible for the patient's therapeutic outcomes. But, as the physician was already being paid for that service, he or she did not appear willing to hand over the earnings opportunity to the pharmacist.

The sole surviving portion of the pharmaceutical care vision is medication therapy management (MTM) where the patient makes an advance appointment with the pharmacist for a private, one-to-one consultation session. The patient brings all medicine bottles and, during a 20- to 45-minute session, the pharmacist ascertains whether all the medications that the patient takes are actually required and, if they are required, that they are administered in the optimal schedule to minimize interactions and adverse drug events (ADEs). Medicare pays for such MTM sessions because there is the possibility that dangerous and costly ADEs might be avoided or prevented, and that the pharmacist might be able to eliminate some of the patient's medicines as being superfluous. Often, patients take 14 or more medications and some of these are to treat side effects caused by the combination of the medications.

Summary and conclusions

The pharmaceutical industry is large, powerful, and wealthy. Whether that description will hold true in the future, with continued globalization of the industry and a partial market shift to Asia and to biotechnology products and genomic therapies, will determine in large part how much the traditional pharmaceutical industry will be able to control and greatly influence pharmaceutical policy in the coming decades.

One may immediately see that the pharmacy profession has had no shortage of internal battles and policy divisions. Many of these have been resolved, often to the chagrin of pharmacists. Also, however, there are an equal or larger number of unresolved issues of a policy nature requiring negotiation or resolution. How these will be resolved if the profession remains split and unaligned can be predicted and should be an unpleasant picture for pharmacists. The pharmacy profession could be more successful in defending its turf if it were to coordinate and work together. Today, there are separate organizations for health system pharmacists, community pharmacy owners, community pharmacists, long-term care consultant pharmacists, managed care pharmacists, pharmacy educators, chain store owners, wholesalers, mail order pharmacies, pharmacy benefit managers, clinical pharmacists, various religious groups of pharmacists, and African–American pharmacists, among still others. Although there are some organizational vehicles for some of these organizations to communicate, the reality is that their respective memberships expect their leaders to defend their particular interests.

The tools for policy analysis are available for the most part and a profession that demonstrates to its patients, customers, or clients that it is looking out for their best interests and long-term welfare can expect to have the powerful voice of the consumer constituency on its side. Would that not be a welcome sight?

Review topics

1 Identify Medicare-related data sources to address cost and quality-of-care issues.
2 Discuss quality-of-care issues relevant to pharmacy practice using the Donabedian approach.
3 Identify any two practice or policy related question relevant to pharmacy practice.

References

Chan L, Houck P, Prela C, MacLehose R (2001). Using Medicare databases for outcomes research in rehabilitative medicine. *Am J Physical Med Rehabil* 80: 474–80.

Civan A, Köksal B (2010). The effect of newer drugs on health spending: do they really increase the costs? *Health Econ* 19: 581–95.

Cockburn IM (2004). The changing structure of the pharmaceutical industry. *Health Affairs (Millwood)* 23(1): 10–22.

Donabedian A (1996). Evaluating the quality of medical care. *Millbank Memorial Fund Q* 44: 166–203.

Ess SM, Schneeweiss S, Szucs TD (2003). European healthcare policies for controlling drug expenditure. *Pharmacoeconomics* 21: 89–103.

Gold M, Siegel J, Russell L, Weinstein M (1996). *Cost-Effectiveness in Health and Medicine.* New York: Oxford University Press, 149–50.

Institute of Medicine (2009). *Initial National Priorities for Comparative Effectiveness Research.* Washington, DC: National Academies Press.

Kilian J, Stubbings J (2007). Medicare Part D: selected issues for pharmacists and beneficiaries in 2007. *J Managed Care Pharm* 13(1): 59–65.

Neuhauser D (2002). Ernest Amory Codman MD. *Qual Saf Health Care* 11: 104, 105.

Riley GF (2009). Administrative and claims records as sources of health care cost data. *Med Care* 47(7suppl1): S51–5.

Scanlon DP, Hendrix TJ (1998). Health plan accreditation: NCQA, JCAHO, or both? *Managed Care Q* 6(4): 52–61.

Seget S (2008). Pharma Trends, 2008–2012. Available at: www.urchpublishing.com/publications/market_trends/pharmaceutical_market_trends_2008_-_2012.html (accessed October 15, 2009).

Smith MC (1991). *Pharmaceutical Marketing: Strategies and Cases.* New York: Haworth.

Winkler S (2007). Policy, law and regulation. In: Fulda T, Wertheimer A (eds), *Handbook of Pharmaceutical Public Policy.* New York: Haworth.

Viswanathan HN, Salmon JW (2000). Accrediting organizations and quality improvement. *Am J Managed Care* 6: 1117–30.

Yabroff KR, Warren JL, Banthin J, *et al.* (2009). Comparison of approaches for estimating prevalence costs of care for cancer patients: what is the impact of data source? *Med Care* 47(7suppl1): S64–9.

Online resources

Agency for Healthcare Research and Quality (AHRQ). "National Quality Measures Clearinghouse." Available at: www.qualitymeasures.ahrq.gov.

Centers for Medicare and Medicaid Services. "Research, Statistics, Data and Systems." Available at: www.cms.hhs.gov/home/rsds.asp.

US National Library of Medicine. "Health Services Research Information Central." Available at: www.nlm.nih.gov/hsrinfo.

Appendix 1

Selected peer-reviewed pharmaceutical practice and policy research journals

American Journal of Epidemiology

Journal URL: www.aje.oxfordjournals.org

Description: publishes empirical research findings, methodological developments in the field of epidemiological research, and opinion pieces.

American Journal of Health-System Pharmacy

Journal URL: www.ajhp.org

Description: provides current information on the clinical use of new drugs and drug therapies based on clinical reviews, research reports, case studies, and commentaries on a wide range of topics in health-system settings.

American Journal of Public Health

Journal URL: www.ajph.org

Description: publishes original work in research, research methods, and program evaluation in the field of public health.

Clinical Therapeutics

Journal URL: www.clinicaltherapeutics.com

Description: includes Pharmaceutical Economics and Health Policy section to address issues in pharmacoeconomic, health outcomes, and contemporary issues related to drug therapy.

Current Medical Research and Opinion

Journal URL: http://informahealthcare.com/loi/cmo
   Description: covers broad range of topics from phase II to phase IV including equivalence, safety and efficacy/effectiveness studies, and pharmacoeconomic outcomes and quality of life studies.

Formulary

Journal URL: http://formularyjournal.modernmedicine.com
   Description: publishes articles related to drug management for managed care organizations and hospitals.

Health Affairs

Journal URL: http://content.healthaffairs.org
   Description: explores health policy issues of current concern in both domestic and international spheres.

Health Economics

Journal URL: www3.interscience.wiley.com/journal/5749/home
   Description: publishes articles on all aspects of health economics: theoretical contributions, empirical studies and analyses of health policy from the economic perspective.

Health Services Research (HSR)

Journal URL: www.hsr.org
   Description: publishes articles reporting the findings of original investigations that expand understanding of the wide-ranging field of healthcare.

International Journal of Pharmacy Practice (the UK)

Journal URL: www.pharmpress.com/ijpp
   Description: publishes original research and review articles on all aspects of medicines management, policy, practice, and education.

Journal of Evaluation in Clinical Practice

Journal URL: www.wiley.com/bw/journal.asp?ref=1356-1294
   Description: focuses on evaluation and development of clinical practice across medicine, nursing, and the allied health professions.

Journal of Managed Care Pharmacy

Journal URL: www.amcp.org/amcp.ark?c=jmcp&sc=current

Description: provides the results of scientific investigation and evaluation of clinical, health, service, and economic outcomes of pharmacy services and pharmaceutical interventions, including formulary management.

Journal of Pharmaceutical Health Services Research

Journal URL: www.pharmpress.com/shop/journals.asp

Description: publishes all aspects of research within the field of health services research that relate to pharmaceuticals.

Journal of Pharmacy & Pharmaceutical Sciences (Canada)

Journal URL: http://ejournals.library.ualberta.ca/index.php/JPPS/index

Description: contains research articles, research reports, technical notes, scientific commentaries, news, views and review articles in the physical, chemical, biological, biotechnical, clinical, and socioeconomic–pharmacoeconomic regulatory aspects of the pharmaceutical sciences.

Journal of Pharmacy Practice and Research (Australia)

Journal URL: www.shpa.org.au

Description: provides peer-reviewed coverage of various aspects of contemporary pharmacy practice and areas of interest and importance to hospital pharmacists.

Journal of the American Pharmacists Association

Journal URL: www.pharmacist.com

Description: provides information on pharmaceutical care, drug therapy, diseases and other health issues, trends in pharmacy practice and therapeutics, informed opinion, and original research.

Medical Care

Journal URL: http://journals.lww.com/lww-medicalcare

Description: publishes articles related to the research, planning, organization, financing, provision, and evaluation of health services.

PharmacoEconomics

Journal URL: www.wiley.com/bw/journal.asp?ref=1098-3015&site=1
Description: publishes economic evaluations of drug therapy and pharmacy services.

Pharmacoepidemiology and Drug Safety

Journal URL: www3.interscience.wiley.com/journal/5669/home
Description: provides an international forum for the communication and evaluation of data, methods, and opinion in the discipline of pharmacoepidemiology.

Pharmacy Practice (Spain)

Journal URL: www.pharmacypractice.org/
Description: its scope includes clinical pharmacy, pharmaceutical care, social pharmacy, pharmacy education, process and outcome research, health promotion and education, health informatics, pharmacoepidemiology, etc.

Pharmacy World and Science (the Netherlands)

Journal URL: www.springer.com/medicine/internal/journal/11096
Description: addresses clinical pharmacy and related practice-oriented subjects in the pharmaceutical sciences.

Quality of Life Research

Journal URL: www.springer.com/medicine/journal/11136
Description: publishes original research, theoretical articles and methodological reports related to the field of quality of life, in all the health sciences.

Research in Social and Administrative Pharmacy

Journal URL: http://journals.elsevierhealth.com/periodicals/rsap
Description: publishes original scientific reports and comprehensive review articles in the social and administrative pharmaceutical sciences.

Statistics in Medicine

Journal URL: www3.interscience.wiley.com/journal/2988/home
Description: focuses on medical statistics and other quantitative methods.

The Annals of Pharmacotherapy

Journal URL: www.theannals.com

Description: publishes evidence-based articles on practice, research, and education to advance pharmacotherapy throughout the world.

Value in Health

Journal URL: www.wiley.com/bw/journal.asp?ref=1098-3015&site=1

Description: reports evaluations of medical technologies including pharmaceuticals, biologics, devices, procedures, and other healthcare interventions.

Appendix 2

Selected pharmaceutical practice and policy research funding sources

Agency for Healthcare Research and Quality (AHRQ)

Agency URL: www.ahrq.gov

Description: The AHRQ supports research to improve the quality, safety, efficiency, and effectiveness of healthcare for all Americans. It funds applied health services research in wide ranging areas including healthcare quality, cost-effectiveness, comparative effectiveness, health information technology, evidence-based prevention, patient safety, and emerging areas in healthcare practice, organization, delivery, and management.

American Association of Colleges of Pharmacy (AACP)

Association URL: www.aacp.org

Description: The AACP is the national organization representing pharmacy education in the United States. It supports new investigator starter grants in basic and applied areas of pharmaceutical sciences.

American Society of Health-System Pharmacists (ASHP) Foundation

Foundation URL: www.ashpfoundation.org

Description: The ASHP Foundation represents pharmacists who practice in hospitals, health maintenance organizations, long-term care, home care, and other components of healthcare systems. The ASHP Foundation supports research to optimize patient outcomes and the design and study of safe and effective medication-use systems.

Commonwealth Fund

Foundation URL: www.commonwealthfund.org

Description: The Commonwealth Fund is a private foundation that aims to promote a high performing healthcare system that achieves better access, improved quality, and greater efficiency. It supports research to improve healthcare practice and policy, especially for society's most vulnerable, including low-income people, the uninsured, minority Americans, young children, and the elderly.

Community Pharmacy Foundation

Foundation URL: www.communitypharmacyfoundation.org

Description: The foundation assists community pharmacy by providing resources for research and development to encourage new capabilities and continuous improvements in the delivery of patient care. It funds projects and research related to medication therapy management, pharmacy administration and economics, patient safety and disease state management.

National Community Pharmacists Association (NCPA) Foundation

Foundation URL: www.ncpafoundation.org

Description: The NCPA Foundation represents pharmacists, pharmacist owners, managers, and employees. The NCPA Foundation supports the growth and advancement of independent community pharmacy. It funds research on critical issues related to independent pharmacy.

National Institutes of Health (NIH)

Agency URL: www.nih.gov

Description: The NIH is the nation's medical agency consisting of 27 components called institutes and centers. It funds grants, cooperative agreements, and contracts that support the advancement of fundamental knowledge in basic medical sciences and applied health services research.

Pharmaceutical Research and Manufacturers of America (PhRMA) Foundation

Foundation URL: www.phrmafoundation.org

Description: The PhRMA Foundation supports training, research, and careers of young pharmaceutical scientists by awarding competitive grants and fellowships. It supports starter grants in the areas of health outcomes, pharmacoeconomics, and patient reported outcomes.

Robert Wood Johnson Foundation (RWJF)

Foundation URL: www.rwjf.org

Description: The foundation supports research that improves the health and healthcare of all Americans. It supports projects that have measurable healthcare impact including service demonstrations, public education, policy analysis, health services research, and evaluations.

Veterans Administration (VA)

Agency URL: www.hsrd.research.va.gov

Description: The VA's Health Services Research & Development Service (VA HSR&D) supports research with respect to organization, delivery, and financing of healthcare, from the perspectives of patients, caregivers, providers, and managers to improve the quality and economy of care.

Index